Air Pollution's the Answer!

Air Pollution's the Answer!

HOW CLEAN AIR POLICY COMPROMISED THE PLANET AND PUBLIC HEALTH

Sarah Schrumpf-Deacon

Nana Janes Books

Copyright © 2021 by Sarah Schrumpf-Deacon

All rights reserved. No part of this book may be reproduced in any manner whatsoever without written permission except in the case of brief quotations embodied in critical articles and reviews.

First Printing, 2021

CONTENTS

The Self-Publishing Experience	vii
Dedication	ix
Forward	xi
Introduction	xiii
1 Past, Present And Future	1
2 Facets Of Earth's Ecosystem	8
3 A Good Look At The Past	15
4 Electricity, Engines And The EPA	18
5 Messing With Mother Nature	28
6 Separating Fact From Observation	36
7 Another Look At Gases	49
8 A Discussion Of Clouds	59
9 Air Pollution Provides Sulfur	68
10 Sulfur And Human Metabolism	79
11 Tweaking Expectations	92
12 History Repeats Itself	103
Sources	111
Informational Graphics	115
About The Author	121

The Self-Publishing Experience

Publication options for first time authors are often limited. Without the backing of established fame or a topic that is unique, mainstream publishers limit their offerings to the tried and true. If it were not for print-on-demand publishing options made available through such companies as Ingram-Spark, this book would never have reached the marker at a reasonable cost. But like all new technologies, the process is not foolproof and there are some considerations every new author should remember before taking the leap into this world.

On the surface, what these companies provide is a promise of an easy online way to take words and turn them into a published work. A simple black and white ebook can be put together with minimal effort but if the author is hoping to have a more polished and professional look to their work, some experience in publishing can be helpful. Actually purchasing and learning the ins and outs of a commercial publishing program may be worth the effort and expense.

With a more modest software approach than standard graphic software, it is also likely that available support may be limited to a printed guidebook instead of actual troubleshooting. Factual details about what makes a good print job may explain what the company is looking for but not always what the client needs to do in order to make that happen. Online and email assistance can be problematic. Some representatives are willing to explain the fine points of the program, others are not. Software settings may not jive with the size or style of the book and the Internet speed can cause delays in uploaded information.

In the case of this book, there were issues with submitted images that met the author's requirements but which triggered alerts and warnings of needed corrections in order to continue. After a great deal of back and forth with staff, it became apparent that the warning was a courtesy message and not an actual determination of poor quality images. Situations like this one can be expected with software that is new or recently upgraded.

While every effort has been made to see that this book is a polished and professional product, some mistakes are expected. At the same time, there is uncertainty about how clear the published photos will be, not because of the quality of he originals but because the author has no understanding of what may happen to the manuscript once it is uploaded. For all its opportunities, self-publishing does have its limitations along with its rewards.

Dedication

To my parents,
who faithfully believed in science
and technology, even though it would
contribute to their deaths

and

To the World
which, if leaders are strong enough
to admit mistakes, might finally escape
sixty years of poor health and
massive drug dependence.

Special Thanks
To Liza L. Peltola, Proofreader, Editor and Advisor
This book would not have been possible
without her patience and support.

Forward

As the idea formed for this book, there was but one thought in this author's mind:

No one will believe this!

And yet, here it is. A science-based version of environmental policy, this account goes against the accepted theories of the day for the purpose of provoking thought and evaluation.

Unlike other books written on the subject, this one focuses on how political agendas, spontaneous events, made to order theories, and culture-driven search engines steered clean air initiatives in the wrong direction. It is not meant to be argumentative or persuasive toward any political viewpoint. In a time of conspiracy theories and misinformation, it tries to point out discrepancies in conclusions--not undercut the value of science.

More than anything, the book hopes to bring out the functional difference between today's efforts and the science of Salk, Edison, Newton and others. Too often, science has become an instrument of persuasion instead of a tool for enlightenment. By presenting the public with information that is contrary to common belief, issues of poor human health and environmental crisis can be viewed as the unintentional consequence of human interaction.

The work's conclusion is based on historic evaluation and basic science with just a twist of creative storytelling. Perhaps it is even shockingly accurate in its practical examination of current perspectives. But, whether the reader agrees with it or not, the time has come to look beyond the present and into the past for the answers that will put the environment back in balance and humans back on the road to good health and longevity.

Introduction

What is Air Pollution?

Ask a hundred people and there will be a hundred and one answers for what constitutes air pollution. While it is easy to describe the exhaust from a vehicle, the smell of the farm down the road or the aroma of the paper mill when the wind changes, an understandable definition of air pollution is far more difficult to come by.

Since the 14th century, the term *pollution* has been used in a mostly cultural manner. It was not until the world moved away from an agrarian lifestyle that defining the term became important. The courts and governments took it upon themselves to define all types of pollution. In simple terms, it was an act or outcome that defaced or contaminated property in a way that made it unusable.

But how does that apply to air pollution?

Explained in detail in later chapters, it is enough to say that what legally constitutes air pollution in the United States varies across each jurisdictions and includes naturally occurring gases that are essential to all forms of life. More importantly, regulating and utilizing them in the same way as oil and coal may have dire consequences for the planet.

To put it simply, many of the gases listed as pollutants by the Environmental Protection Agency (EPA) serve the planet in beneficial ways. Alarmingly, since government has put limits on the natural production of these gases, the health of US citizens has declined. As of 2019, healthcare expenditures made up 17.7% of the national GDP (gross domestic product) according to the Center for Medicaid and Medical Services (www.cms.gov) compared to a fraction of that before such legislation existed.

In the chapters that follow, this book will make a case for revisiting clean air mandates put into place some fifty years ago. It will also challenge the popular "greenhouse gas" theories on which so many computer models predict a doomed Earth. In the end, the direct link between restricting essential gases and the phenomenon of climate change will be made clear.

The Climate Change Controversy

The year 2020 was eventful - for more reasons than could have ever been imagined. A devastating world-wide pandemic, which has not been seen since 1918, and a season of record-breaking weather events would have been enough to push climate change to the forefront. With the first three months of 2021 bringing unusual nationwide winter storms, the discussion has understandably reached a fever pitch.

Calls for quick and decisive action come from celebrities, politicians and activists. But hasn't the world been here before? Is reactive legislation what is needed or will it do unintentional harm?

As a higher species, man believes, and is even taught, that his status gives him dominion over the Earth. He is the planet's protector and strives to do what is best for it. Instinctively, his concern is shaped by his own needs and beliefs. It is this human perspective that confuses any climate change discussion and makes it a difficult and politically charged landmine.

The world agrees that something has changed. Physical and mental health as well as weather and food supplies are drastically different. Could all this be a result of a handful of simultaneous events that clouded human perception? Could science's heavy use of computer modeling have hidden connections that human observation and teamwork would have solved years ago? Is this a weather problem or the breakdown of a simple, well-defined scientific principle?

All these questions are explored as this book makes its case against zero tolerance air pollution policies.

1

Past, Present and Future

Throughout history, accurate information, or *intel*, not only had political value but was essential to survival. Knowing **how** things worked was not important. Being able to *anticipate* and *predict* them was. A steady rain on a warm night might be the difference between hard, seedy nubs and luscious, shiny blackberries. A person who paid attention to those details enjoyed cobbler for dinner or jam through the winter. To miss that connection meant well-fed wildlife instead.

As often happens with social activism, individuals champion only what they see through the lens of their own background. How their vision is viewed by others never registers. A computer whiz sits comfortably at his desk and recommends stricter environmental measures based on computer models that focus on one isolated theory. The farmer in the Plains worries that he will not make a crop this year because the weather wavers between long dry spells and heavy downpours which hinder growth. Both individuals see a problem--just not the same problem. For the farmer, the problem is now. For the activist, the problem is the future. To the media, one is science and the other an act of God.

Recycling Symbol

To further complicate any political discussion, environmental advocacy has morphed into a cause that is marketed and sold. Fact and impression differ. Embracing foods, vehicles and low water-use shower heads as a way to help the planet has everything to do with personal feelings and nothing to do with proven environmental science.

In contrast, Earth's long history tells us that the environment is anything but helpless.

Encompassing a huge network of interconnecting cycles, an ecosystem cannot be reduced to simple data or trends. Without a partnership that includes intuition and application, the chance of error comes from accepting incorrect theories that do not work in the real world.

Past and Present Research

Before computers, scientific research was tedious and time consuming. While it still can be, the process is faster when technology is used. Unintentionally, the used of technology can cause distance between process and the findings that warps the results. Small discrepancies that would trigger a flash of human understanding can be lost.

In the past, good research required multiple experiments, detailed notes and large blocks of time by one person. The scientist participated in every aspect of the study. Today's use of multiple choice surveys, literature reviews and analysis of non-standardized records cannot compare to the slow and collaborative nature of research done in real time.

Still, the basis of all quality research begins with the **Scientific Method.** First documented by Sir Frances Bacon some 400 years ago, the method has been used for much longer. Although many versions exist, it is a structured, step-by-step problem-solving technique that can be applied to any question. These are the typical steps that are followed during an experiment.

1. **Observe** (Sometimes listed as *Background Research* or *Identify the Problem*)
2. **State the Hypothesis** (Make a detailed guess why something happens)
3. **Plan** (Design a way to see if the guess is accurate)
4. **Act** (Carry out the plan, hold the experiment)
5. **Record the Results** (Write down everything that happens)
6. **Evaluate** the Results
7. **Repeat** (Until the results can be repeated several times in the same manner and with the same results, the hypothesis is not considered true)

Scientists, through time, deserve respect for their accomplishments, but that respect should not automatically cancel out scrutiny of their work. Blind acceptance of a theory based on being the first or being the most published goes against the highest principle of the Scientific Method. Until such time as a theory can be proven and repeated by a community of scientists, the findings are not considered accurate.

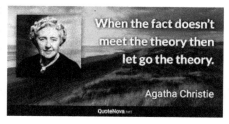

As will be explored in the next section, the computer and particularly the Internet, changed the way information was handled. Theories found on the Internet were not always based on fact but were accepted as reliable simply because they found their way into print. Even a writer of mysteries understands that fact is what matters, not supposition.

Computers Replacing Thought

Today, information is a business. At the same time that devices opened up the world for many, computers and search engines have narrowed human thought in unexpected ways. Initially planned as a means of accessing varied viewpoints and escaping the image of bias, in reality, this new technology was a flawed tool in the hands of those who knew little about its limitations. Scientific research, based solely on computer analysis, runs the risk of being ***accurately wrong***. A confusing combination of words, ***accurately wrong*** perfectly describes the process by which clean air policies were developed.

Based almost solely on the dangerous smog (i.e. the words smoke and fog) events in large cities like Los Angeles and New York in the 1950s and 1960s, air pollution was a rare if not completely unknown experience for most of the country. As the situation in metropolitan areas grew, even a light haze was attributed to polluted air.

Determining a method of smog control could have followed a path something like this.

Instead of starting with the question *What is air pollution and what causes it?* the beginning of clean air policy might have been *How can air pollution be reduced?* By considered only *how* instead *what happens if air pollution is reduced?* scientists only needed to be concerned with one small part of the an otherwise huge problem. The Will Rogers' definition of an expert--"a man fifty miles from home with a briefcase"--was a common yardstick validating outcomes in the early days of environmental activism. One person offering an outside opinion was all that was needed to justify a decision.

The first step would have been to determine what was in the atmosphere during a typical smog event. The next step would have been to determine which activities could impact gas emissions. This information would have been put into an old key punch style data processor and the program would have determined the most likely options for reducing the target gases. A timeline for implementation would have been created with penalties for failing to comply.

Here is where *accurately wrong* decisions were made.

Researchers accurately determined the gases present in a smog event. They accurately determined which activities produced these gases. The data processor accurately analyzed the information based on the criteria of reducing air pollution and a workable schedule for implementing the guidelines was created.

So how could something so accurately applied have gone so wrong?

Often called **unintended consequences**, unexpected results often happen when quick decisions are made without adequate thought. Clean air policies have avoided tough topics such as overpopulation, building size which alters air circulation, and an atmosphere that has more functions than only something to breathe. By simplifying the problem into a pile of data, the chance of a long-term, accurate and sustainable answer was always questionable.

If this scenario is close to accurate, the potential for wrong turns was high. Researchers were wrong to label multiple gases within a smog event as pollutants without individual testing. They were wrong to use the largest cities as a benchmark for broad national restrictions since cities had additional contributing factors that most of the country did not. They were wrong to trust data without allowing it to be scrutinized by other disciplines over time. Most of all, they were wrong to set a zero tolerance standard without understanding how each individual gas worked within the environment. The unintended consequence of this legislation was a handicapped environment that had to respond in kind to keep the world in balance. That response is what is now called climate change.

Technology: A Cautious Love Affair

There is no going back to a time when technology does not factor into daily life. Even after forty years, this digital love affair has yet to mature into a strong reliable bond. Computers are simply not capable of the subtle thought that comes with human observation. Raised on the image of Star Trek's main computer that never misunderstood a question or failed to have an answer, current users believe in an image that is little more than fantasy.

An Example of Junk Science

Much of scientific discourse today relies on statistical analysis. This book could resort to such tactics but will try to avoid them. To give an example of how such claims are made, here is one *accurately wrong* statement that supports the premise that clean air policies have gone too far and are endangering human health as well as the environment.

> *According to IQ Air (www.iqair.com), the United States is ranked 84th out of 106 countries for most polluted air in the world. In other words, 83 countries have more air pollution than the United States and only 22 countries have less air pollution than the US as of 2019. At the same time, Peterson-KFF Health System Tracker (www.healthsystemtracker.org) ranks the United States as having the highest per capita health care costs on the planet. The assumption is that as air pollution has gone down, health care costs have gone up meaning air pollution may not be as detrimental to health as once believed.*

On the surface the information sounds accurate and reasonable. A look at ownership of the sites puts the information into a different light. Each site is independently owned and on some level supported through non-profit funding. While this point is not meant to cast doubts as to the ethical nature of the organization, simply put, each website is under no legal obligation to insure the accuracy or standardization** of its information.

Other factors that make the information unreliable include different measures across different countries, economic limitations, political policy and human error. Even if a correlation between a reduction in air pollution and an increase in poor human health exists, basing it on this type of information makes it a flawed assessment.

This book intends to draw a clear connection between atmospheric changes over the last fifty years and the rise of poor health in the United States. It will do so by walking the reader through a century and a half of history on which current climate change theory is based. Then it will provide details as to how less air pollution has impacted the food supply and human health.

In the end, it will be the readers who decide if the author has made the case for rolling back decades of harsh environmental policy. It will be the audience's choice as to whether a handful of unsustainable urban locations have the right to put rural communities at risk in order to maintain the status quo.

**Standardize is a term used when comparing numbers. The question is do the numbers represent similar measures. Think of apples. Think of all the ways apples can be different—size, color, weight, sweetness, etc. Each locality, be it state or country, has a different measure and definition for air pollution. Comparing numbers without standardizing the measures makes the comparison *accurately wrong*.

2

Facets of Earth's Ecosystem

Compared to other sciences--Chemistry and Astronomy, for example, Environmental Science must consider human influences in its theories. In contrast to the rigid and predicable outcomes common in the physical sciences, the study of the environment is where all sciences, physical and social, meet.

This work will begin with the physical and then incorporate the social sciences. The starting point is a few thousand years ago when mankind had only observation and trial and error as scientific tools.

Known as Alchemy in its first days, scientific discovery came when the belief in magic meet physical reality. Categorizing everything seen or unseen as Earth (land), Water, Air and Fire (energy), this beginning would eventually become the study of modern science.

The Overworked Land

For all of recorded history, the land has provided a solid and usable foundation on which civilization built its cities. It provided the resources for society to transition from survival to a level of comfort that no one could have imagined those many years ago.

While its mountains, valleys and deserts have changed little, *how* land is used is very different. Gone are many of the regional identifiers that made global ecosystems varied and unique. Everything man had was dependent on where he lived. Homes of logs were common in heavily forested countries but those made in drier climates were adobe or grass. Rich differences in cuisine came about as they relied on what could be grown and harvested naturally, not purchased from an international food market. Above all, the earth beneath human feet was respected because on it, an individual staked his success.

Current scientific advances make it possible to force resources to work in ways that are environmentally incompatible. Crude oil is used to make fabric and food containers instead of warming homes and protecting materials from water and sun damage. With public perception leaning towards plants as the answer to climate change, woodlands become ghost forests when trees die of old age and rot without purpose. Timber stands that do not die out fuel the next wildfire instead of building long lasting furniture, fences and homes. Diamonds are manufactured because finding them is politically problematic while natural windbreaks are destroyed to build smooth roads, large airports and challenging golf courses. The question is what did it cost the environment for human convenience?

The line between protection and mastering the ecosystem has blurred. Science no longer concerns itself with regional ecological health. The environment must conform to best practices, even when theory does not work in the real world. Raw materials produced in one region are readily consumed by another. Sustainability is an abstract catch phrase that means maintaining the status quo rather than being ecologically accountable.

Mixed Messages for Water

Like dry land, the size, shape and location of the Earth's bodies of water have not changed much over the last thousand years. That said, modern civilization has something of a mixed attitude toward water.

How water is perceived relates to its location. Fresh water (drinking) is managed. Dams are built to support communities that would otherwise be left thirsty. Communities which settled too close to water keep it away by building levees. Wells are dug to provide water in isolated locations but rain is diverted into ditches and storm drains because the soil that would naturally absorb it and replenish underground reservoirs has been covered with asphalt and concrete. As for travel, if the world did not provide an easy route between one body of water and another, civilization built canals with hard labor and shovels. Even when the environment has other plans, human intervention channels water whenever and wherever it is needed.

In contrast to fresh water, other sources are honored. Unique ocean species are objects to study or photograph. Ice floes and glaciers are measured as if they were children meeting certain milestones. The ocean's once valuable purpose as a source of renewable food, energy and raw materials takes a back seat to preservation efforts. Even when death comes naturally, creatures are mourned by activists as if all death is unnatural and preventable.

No matter how it is perceived, water has a specific and integral place in Earth's ecosystem. Known as the *universal solvent*, the same water that makes a great glass of fresh lemonade has the power to wear away mountains and clean smoke-filled air with an afternoon rain shower. But fulfilling its purpose depends on it being used, not conserved. As environmentalists treat water like a savings account, the practice of conserving water disrupts the global flow of water and minerals on which life depends (see Chapter 8).

Rainbow Over Ocean

Atmosphere on a Diet

Taken for granted more than Earth, Water and Fire, the atmosphere's invisibility hides its contributions very well. Used by all life forms in some way, it is what sets Earth apart from every other identified planet and moon. Without it, life would cease to exist. Treating air like clothes that have been worn too many days in a row, people only notice it when it smells or looks unhealthy. Unlike clothes that require washing to clean them, the atmosphere has an effective and automatic system called "weather" to do the job.

Not just something to breathe, Earth's atmosphere is the force that balances the scales between urban and rural areas. From partnering with plants to whisk away pollen for fruit and vegetable production to becoming a buddy with the sea for aerating underwater ecosystems, the air takes what some call pollution and puts it to work. Dismissed because it offers nothing tangible, the atmosphere secretly holds the key to a working environment that operates smoothly.

Today, the atmosphere is an anorexic version of what it was in the 1950s. Unlike today which limits the composition of air to just five components (nitrogen, oxygen, carbon dioxide, water vapor and dust), as much as 10% of the atmosphere was considered a diverse and important combination of gases. As science found ways to separate each individual gases from oxygen and nitrogen, manufacturing found ways to use them. Helium, neon and argon were synthesized by industrialists who had no qualms about putting the atmosphere on a diet. Now, carbon dioxide levels have activists sounding the alarm. With *Clean Air* legislation further restricting the production of several naturally occurring gases, the atmosphere may look lean and healthy but it lacks the muscle to perform its environmental duties.

Trial by Fire

For early civilizations, fire was the change maker. Believed to be part of the sun, its heat changed everything. Solids changed into liquids which then became gases. Without fire, food was just animal flesh and raw plants. Rocks could not be melted down into jewelry and tools. Water could not be boiled to make warm beverages or prevent food borne illness.

Of these four, fire was believed to be the most powerful. Now, civilization worships energy instead of fire. Its need for a constant and reliable power source drives the discussion of climate change. As modern civilization searches for a low-maintenance, always available supply of energy, the question becomes at what cost to the environment will man have his way?

Fossil fuels and renewable energy are still available in every corner of the world. The environment provides multiple sources of fuel and each of them proves beneficial to the environment as well as to civilized life. Regrettably, these natural sources of energy do not seem to suit the concept of environmental protection. As activists promote sustainability, they seek to herd communities away from the very fuels which are the most renewable. Oil, coal, wood and gas may seem environmentally unhealthy but that assumption may not hold up as this narrative moves forward.

The Mythology of the *Carbon Footprint*

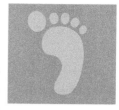

A particularly challenging aspect of writing this type of book comes when new information challenges old beliefs. Without disrespect for the old, an author must find a way to explain what amounts to a misunderstanding of scientific principle. One such example is the very common idea of a *carbon footprint*.

How this concept came into being is not completely known but it apparently started with the term *environmental footprint* which was used in the mid-1990s as a part of city planning and economic development. Used to draw attention to evaluating decisions based on their environmental impact, it focused planning on trash disposal, water purification, sewage treatment and yes, air pollution. Fundamentally different from sustainability, the concept is very applicable to the discussion of environmental health as it acknowledges a community's responsibility to handle its own mess.

From there, how it became *carbon footprint* is something of a mystery. Perhaps it was misquoted in an article about greenhouse gas theory. The possibility also exists that in identifying carbon dioxide as a factor, there was confusion that carbon was a detriment to the environment. What ever the origin of the term, *how* it is applied to climate change and air pollution is flawed in its understanding of environmental sustainability. With a bit of history and some added science, it might be possible to bridge the gap between old and new understandings.

In 1789, Antoine Lavoisier proposed the Law of Conservation of Matter (Mass). In that law, he stated that matter could neither be created nor destroyed. In other words, the world has a set number of carbon atoms. Those atoms can be combined to form a wide array of products and life forms but once those atoms are gone there is no more. The wonder of this system is that it is always recycling its resources. Carbon-based gases are instrumental in this process.

Today, carbon may be in the form of carbon dioxide. Tomorrow, a plant make it into the sugar and fiber in the piece of fresh fruit that a child eats for a snack. That newborn baby next door is an estimated 18.5% carbon and the diamonds that are a girl's best friend are 99.5% pure carbon. Even the trash that was dumped in the ocean or buried deep underground will eventually decompose and resurface as methane, natural gas or crude petroleum. But, it all starts with carbon in the atmosphere.

What cannot be ignored is the pervasive nature of this ecological cycle. Should climatologists recommend reducing carbon to its solid form, it is likely that the environment would respond in two ways. First, it would cause a reduction in plant life across the world. Second, geological events would occur that would send thousands of cubic feet of ash, carbon dioxide, and other carbon based gases into the atmosphere to make up for what has been lost. The environment will balance the scales even if the perception of modern science believes carbon dioxide is a danger.

One last consideration as to why the term has become so popular is encouraging customer loyalty for products believed to be good for the environment. Like so much of what has become "scientific research" in an capitalistic world, studies are often produced to support marketing claims that increase sales. Ultimately, doing away with carbon in the atmosphere hampers its smooth operation and increases the chances that climate change will get worse.

3

A Good Look at the Past

Why is history so often treated as little more than an engaging storybook? One reason may be that the rapid changes of the 20th century disguised its relevance to current affairs. Taught, too often, as compartmentalized dates and events without continuity, history seems to have little to connect current environmental problems with the past--that is until one looks beyond the details and studies the way events occurred in concert. When seen together, it is easier to see present day mistakes when reflections of the past show a pattern.

As new accounts of historical events become available, the perception of those events changes. Without a look to the past, current leaders run the risk of confusing quick decisions with a well-planned course of action. Quick decisions lead to unintended consequences and it is those many unintended consequences that are the focus of this book.

Back to the Beginning

For thousands of years, man thrived, not only because of his ability to think and use his opposing thumbs but, by using fire on a daily basis. The fuel sources may have changed with the location and time period, but as far as the environment was concerned, it worked well.

As populations grew, fire became increasingly unsafe. Accidents led to the devastation of whole cities. For all of its benefits, open flame produced unwelcome by-products.

Great Chicago Fire

With man's growing knowledge of how the world worked, the time came for a safer and more efficient energy source. Inventors considered many alternatives. Wood, coal, whale oil and other fuels were replaced by natural gas and oil. Other inventors designed new ways to use fossil fuel energy that would change civilization until it was almost unrecognizable. But how these inventions impacted the ecosystem may factor into the discussion of climate change more than modern scientists realize.

Progress Interrupted by Conflict

Contrasting with the scientific enlightenment of the period, the 1800s were not a time of peace. The United States had its Civil and Indian Wars but the world as a whole had its share of skirmishes. Hundreds of conflicts, from small revolutions to outright battles, were fought during the middle of the 19th century. The cause of these conflicts varied but many evolved from a simple desire to control the environment. Land boundaries, resources and trade routes as well as water rights were enough to bring opposing sides to violence.

Through it all, inventors kept at their work. Many of the items found in today's households had their beginnings in the 1800s. Appliances such as the waffle iron, typewriter, sewing machine and toaster added to the convenience of daily life. All of them had one problem. They relied on burning fuel or human effort as a source of power.

It would be the end of the 19th century before conflicts in the world would calm enough for a major discovery to become workable. Civilization would get its convenient, always available source of energy in the form of electric current. While electricity would change daily life forever, it would slowly unintentionally change the way the environment worked. To this day, electricity is seen as a safe and clean energy source that is environmentally friendly. Science may tell a different story.

4

Electricity, Engines and the EPA

Understanding history alone is not enough to keep civilization from making some rather harsh mistakes. In a 2017 article entitled, ***Knowing Just Enough to Be Dangerous***, the Dunning–Kruger effect is explained and demonstrated (www.psychologytoday.com). According to the article, those who suffer from this human failing display overconfidence and limited critical thinking skills when given enough information to understand the basics of a concept but not its complexity. A common trait found in some journalists, celebrities, elected officials and administrators, these persuasive people sideline those with real life experience while spreading misleading and incomplete information.

In the next few chapters, examples will be provided as to how overconfidence and a lack of self-awareness about the environment came together to put the planet on the path of climate change. Especially in times of crisis, the Dunning-Kruger effect has the potential to make matters worse instead of better.

Electricity Changes the World

Initially produced only through chemical reactions in batteries, electricity seemed destined to be a discovery that would never be more than an oddity with limited practical application. With the Civil War at an end, the discovery suddenly became workable on a larger scale as electric current. Many inventors contributed to its discovery, but it was Thomas Edison that made it workable on a global basis.

After multiple cities across the country burned as a result of installing gas lights throughout downtown areas, Edison proposed adapting existing infrastructure to carry electric current to every building. With miles of copper tubing which led from a central substation to most locations, Edison easily installed wiring that provided light to hundreds of homes at a time. Use of electricity for other uses would have to wait for Harvey Hubbell to design the electric pull-chain socket (1896) and the electric socket and plug combination (1904) before electricity could be used for appliances that previously existed in battery form. By using existing infrastructure, electric current could flow through wires in a convenient and cheap manner. Using a water wheel adaptation that would be called a *turbine*, huge amounts of continuous power were produced with very little effort. Unlike gas, coal, oil or wood, electricity could be produced miles away from where it was used. The turbines which would produce the current operated on water power as well as steam power from coal or oil fired boilers. The versatile energy source that civilization wanted was finally available.

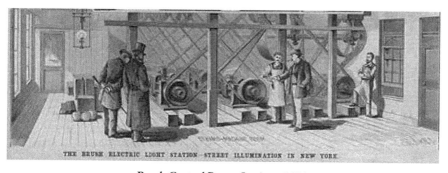

Brush Central Power Station - 1881

By the early 1900s, the majority of large urban areas used electricity for general lighting and other uses. Cities such as San Francisco (1876), Manhattan (1882)and San Diego (1888) were some of the first to make the switch from gas and coal oil to electricity. Even rural areas across the globe had some form of electricity by the 1930s.

Hailed as a wonderful source of clean and safe energy, electricity had all the positives of burning fuel and few of the problems. By recycling infrastructure from gas use and focusing on converting existing mills to hydroelectric plants along major water ways, the only investment needed to bring electricity online was running the wires. In a matter of decades, electricity replaced fossil fuels and became the primary source of energy for the modern world.

But overconfidence in this new energy source and a lack of environmental awareness factored heavily into what may have been the first step toward a world of *climate change*.

At the same time as electricity grew in popularity, the study of meteorology was beginning. It is no surprise that there is no hard evidence for the conclusion that a world without air pollution would be the precursor of a major shift in unfriendly weather patterns around the world. There is no doubt though that by 1930, the USA experienced a years-long drought which would change agriculture forever.

Dust Storm Texas
1935

The Dust Bowl period in history (1930-1936) was chalked up to poor farming practices, bad government policies and a moody Mother Nature. Too busy surviving a global economic depression and military conflicts to give much thought to the *why*, science would not connect the growth of the transportation industry and its increase in noxious fumes with a return of stable and predictable weather patterns in the 1940s and 1950s.

Today, the same pattern is repeating itself in the Western United States. Those who have a little knowledge and great influence beat the drums of panic to get support for massive infrastructure changes that are based on personal perspective and political loyalty, not scientific fact.

In a New York Times interview transcript posted on June 1, 2021, former President Barrack Obama speaks to how the current administration will finish "our climate change agenda." The purpose of the interview was to promote yet another book that passes off his personal perspective for well-researched knowledge and real-world experience.

Implying that "if we don't get a handle on climate change," there may be no one around to judge its failure. the former president attempts to influence acceptance of this path through fear and not science. Without citing a single scientific principle or plan but he pledges to bring about a "clean economy" if trillions of dollars are spent on solar and other unproven theories. If electricity was the beginning of the climate change crisis, what would a global conversion to similar energy sources cost the environment? With just enough information to be dangerous, could the former president be recommending the tools that would led to the human extinction he predicts?

Like many others in positions of authority today, former President Obama gives opinion and misinformation without direct knowledge of what might be environmentally wrong or how such action will solve a, as yet to be identified, cause. This interview is a prime example of a person displaying Dunning-Kruger effect.

The Internal Combustion Engine (ICE)

Mention the word "engine" and an image of a car or train might come to mind. Today, engines and transportation go hand in hand, but that was not always the case. What is now called an *engine* did not start out as a means of transportation.

The first ICE was created in the mountainous regions of Southeast Asia. Referred to now as a "fire piston," the design would be the basis of the diesel engine some two thousand years later.

Adapted and revised to meet different needs, the engine provided far more benefits than just making a fast-food run. Some were motion driven by people or animals. Others were powered by a wide range of fuels such as gun powder, steam, gasoline and even electric batteries.

One of the most popular in the 19th century, the steam engine was used to pump water, grind grains and power machinery such as looms and river boats. It would be the 1880s before inventors would be able to redesign the steam engine to make it portable enough for vehicular travel and capable of using gasoline for fuel. Even then, global availability of cars was limited to a few thousand a year and their use was limited by the availability of refined gasoline. It was not until the 1930s that family vehicles became commonplace in America.

It seems quite the coincidence that as automobile use exploded in the 1930s and 1940s, the droughts that had devastated farming communities throughout the Plains and Southwest United States abated. What had been called an act of God. might have had more to do with Mother Nature finally having the air pollutants she needed to bring rain back to the plains rather than prayers and wishes. (See Chapter 8)

Of all the negatives that are attributed to the gasoline powered engine, there is one element of its operation that deserves such criticism.

In the early days of the gasoline engine, the "knocking" noise made by the engine was distracting at best. The addition of lead to gasoline solved this problem but had the unintended consequence of producing a new noxious material to the exhaust. There is no doubt that the resulting lead oxide was both dangerous to life and dangerous to the environment. The introduction of unleaded gas was controversial and was accompanied by other additives and higher production costs. When unleaded gasoline successfully eliminated one pollutant, ethanol was advocated to bring other pollutant levels down. Promoted as good for farmers and the environment, its manufacture was only cost effective if gasoline prices soared. Today, unleaded gasoline with up to 10% added ethanol is sold to the public at a cost of roughly five times what gasoline cost in the late 1970s.

Could there really be a direct connection between air pollution and stable weather patterns? Could the use of electricity as a primary energy source be the cause of the world's environmental difficulties? Only with specific and detailed research can those questions be answered. In the end it will be up to the reader and the scientific community to decide if these deductions are accurate.

Selden Model 25

The EPA: Influencing Environmental Policy

Of all the pieces of environmental legislation that have been passed by the United States Congress, the Clean Air Act of 1970 has perhaps been the most influential in bringing the American people to the place of climate change. Interestingly, it was that piece of legislation that created the agency that is now known as the Environmental Protection Agency(EPA).

Faced with a long string of environmental disasters and events in the months before the summer of 1970, then President Richard M. Nixon addressed a joint session of Congress on July 9th of that year. In that speech, Nixon proposed a 37-point shift in environmental policy that would set a new standard for monitoring and protecting the environment. While what was understood about the environment at that day was a hodgepodge of theories and conflicting perspectives, this agency may have been one more governmental decision based less on science and more on political pressure and popular opinion.

Part of Nixon's environmental laundry list included setting up a new agency to handle all environmental issues. The new agency was created by presidential executive order and was funded and operational within five months of Nixon's speech. Given broad powers to monitor, research and regulate all activity deemed environmental, the agency answered to the Office of the President and was given authority to determine the success or failure of its own programs without outside scrutiny (www.epa.gov).

Certainly needed to bring industrial facilities to heel and mandate the restoration of contaminated land and water, the EPA and its successes have not had a significant impact on climate change. The fact that the world's climate has become more erratic since the adoption of these measures should have been a clue that environmental policy was headed in the wrong direction. Life would, instead, provide political scandals, Middle East Conflicts, the first major tax cut to the wealth since the Depression and the death of Elvis-the King tto fill the news cycle in those first fifteen years of adoption.

At a glance, the agency's quick formation and the lack of qualified oversight should have been red flags to anyone in Congress. A harder look at what transpired prior to its formation gives a clue as to how a powerful agency that should have been founded on scientific principle was allowed to support a political agenda of a sitting president and the agendas of those that followed.

The list of unintended consequences starts with the 37-point to-do list laid out by Nixon. In it, with no apparent scientific basis, he clearly targets the automobile industry, underwater dumping of waste, and the transportation of oil by ship. Most, if not all, of these points were ongoing concerns for his home state of California. Known for federal lobbying to gain a larger share of environmental resources, California's influence into environmental policy has become great.

In today's political community, California holds a total of fifty-five seats (two Senators, fifty-three Representatives) which is more than 10% of the total legislative body. With the additional influence of the Speaker of the House of Representatives (2021) and the Vice President of the United States (2021) as California natives, small rural districts rightfully question if US environmental policy is one that benefits all states or predominately California.

Once in operation the EPA was limited by the wording of the agency's mission. In reality, the agency's scope could only target practices which were considered harmful to the environment. Research paid for by EPA and private grants typically supported existing policy and not information that contradicted it. In a review of EPA documentation which tracks pollution levels, the reality is in direct contradiction to the findings of many studies carried out between 1985 and the present.

Documented numbers shows a drastic drop in air pollution levels across all categories over the first fifteen years of regulation. Studies completed after 1985 that conluded that air pollution was to blame for countless health conditions would appear to be based on misinformation. Instead of being an affirmation of air pollution's negative effects on health, these studies may have instead supported a growing medical industry and national health care agenda.

For fifty years, the country has been subjected to a roller coaster of environmental anxiety as every research study claimed a new environmental danger. Were all these gases and chemicals actually dangerous or was confirming EPA policy the only way to get funding for modern-day research?

Today, the EPA's mission seems more aligned with political policy than environmental protection. Two recent examples come to mind.

In January 2020, during the Trump Administration, the herbicide Glyphosate (a chemical in Roundup and other weed killers) was retested by the EPA and found environmentally safe. This second test took place after a federal judge ruled that evidence supported the link between regular use of the herbicide and the development of non-Hodgkin's lymphoma. A second retesting situation occurred as the pandemic took hold. In early July 2020, the Trump Administration announced EPA approval of two Lysol cleaners for control of COVID-19 infection. While the EPA is tasked with determining the environmental safety of products, it has not typically crossed over into approval or recommendation of products. At the time of this announcement, the EPA press release did not add a list of all the cleaning products for which safety testing had been done, nor did it refer the public to the CDC list of cleaning products which were recommended.

As the EPA reports its success in reducing emissions and partnering with businesses to continue to limit carbon-based gases, it seems unlikely that the agency will take a chance and research alternative climate change theories. Forcing private initiatives into getting with the program or risk losing government contracts and political support flies in the face of true scientific inquiry.

The Bill and Melinda Gates Foundation as well as Jeff Bezos and Elon Musk have all committed millions of dollars in grants to the study of climate change. But like all grants, these funds are heavily restricted and must fall within the parameters set by the grantee. Private funds are no different than state and federal grants. Each payout begins by setting the desired outcome, not a goal of finding the truth.

Recently, major news outlets reported that an EPA report linking human interaction to climate change had been withheld during the previous administration. The report does not seem to include any admission of culpability for the agency itself. This book will look at the history and science that may not have been considered as the EPA was so hastily founded.

Historical Note:
By the early 1900s, California had become exceedingly important as a port for trade and source of gold and other minerals. Unfortunately, its population quickly overwhelmed the sustainability of the typically dry coastal climate. With an average annual rainfall of between 10 and 25 inches depending on the location, Southern California has a been able to meet its power, water and food needs by diverting resources from other regions and states. Without electricity from the Hoover Dam on the Colorado River, diverting water for irrigation from the same river and claiming legal authority to mountain runoff from the Sierra Nevadas mountain range, the cities of San Diego, Los Angeles, and San Francisco would fail to meet the needs of its citizens. Even today as drought again grips the region, California is declaring a "water crisis" within its own borders.

5

Messing with Mother Nature

During the mid-1970s, soft tub margarine hit the market for the first time. To a curious public, it was an *amazing* product. Consumers did not ask what was in the butter-like spread that kept it soft even in the refrigerator. It was enough that it spread easily without damaging the bread. Like the introduction of this new product, the public seems more interested in accepting the novel rather than the accurate when searching for an answer to climate change.

As Chiffon margarine came on the market, one humorous advertiser promoted it by **fooling** Mother Nature into thinking it was nature's butter. Each new version of the commercial had an unsuspecting main character praising the product as "her butter." When the unhappy main character realizes she has been fooled, the marketing concept ends with the now iconic statement, "It's not nice to fool Mother Nature!"

Chiffon Margarine Ad (Mid 1970s)

Like the margarine, the public's view of the environment may be something it is not. The basic principles of Environmental Science have not changed, but how the environment is perceived has. The result of a good advertising campaign, what is now reported as discovery may be nothing more than a broad-based intellectual community looking to promote its program and brand.

In order to recognize what climate change is, understanding how the country, and world, got here helps. Once the human factors are understood, the environmental ones become clearer.

A Shift Away from Tradition

By the 1960s, after four decades of economic struggles and wars that ended with a half a million deaths in the United States alone, both the US and world were tired. There needed to be more to life than working, saving and making-do. Moderation, which had seen the previous generations through hard times, was not a popular concept. Life needed to be embraced and enjoyed.

In frustration, young people protested traditional restrictions and pushed a philosophy of equality and fairness that was as much about *having it all* as accepting individual differences. Raised to save for a rainy day but wanting a different type of lifestyle, workers threw caution to the wind and went after the new and different.

Believing that the environment and its weather were randomly occurring factors out of the human control, grasping nationwide prosperity seemed attainable the first time in a generation. The transition away from conserving resources set in motion conditions that had not previously existed in the environment except in isolated locations.

Use and Reuse was No More

A sudden increase in population followed WWII and with easier access to credit consumerism became the norm. Gone were the days of small, serviceable wardrobes, old houses and the one-car family. Convenience was what mattered. Low-cost products changed the definition of wealth. Women went to work, not only to realize their own dreams but to allow the family to afford goods considered a luxury as little as a decade before. Equality meant equal buying power as much as equal rights under the law.

Environmentally there were problems. The new preference for cheap and temporary goods meant ever growing volumes of trash. Even with thrift stores and donation centers, small communities found themselves purchasing large tracts of land for nothing more than waste disposal. Politically, it was safer to encourage waste than risk an economic slowdown.

City planning also changed. Homes were built quickly and cheaply without thought to building wealth. In the years that followed WWII, tract housing projects created an estimated six million small single family dwellings in California alone. Buying *on time* would get returning GIs a house, car, television set or living room furniture that did not always last as long as the payments. With a car in every driveway and miles of roads connecting these growing communities, the combination of close quarters and inefficient engines was too much for the environment to handle.

The ocean breezes, pastoral settings and clear skies that drew people to suburbia, soon clouded over with exhaust. Unlimited space, unlimited resources and unlimited mobility put a strain on the environment that no one could have anticipated.

Feelings Overwhelmed Intellect

The wars (WWII, Korea, Vietnam) that followed the Great Depression had been a unique opportunity for surgeons and medical hospitals. With the aid of new drugs like penicillin and sulfa to fight infection, health care took on a different meaning on the battlefield. A fine line existed between following protocol and taking a chance at saving a life. A new treatment that worked was passed on to others without years of trials and study. Any person that was sent home alive meant a job well done regardless of the impact on quality of life. Not since the Civil War had so much progress been made in dealing with otherwise life-threatening injuries.

Medicine became about keeping an individual alive, not always about sustaining health. With vaccines for childhood diseases, a push for better nutrition for the young, and advances in surgery and pharmaceuticals, the world's population would triple in a little more than three decades.

Environmentally, it took the World by surprise. Even when population controls were recommended, the number of inhabitants continued to grow. Food became a global commodity and hunger became a global problem. With scientific advances, mankind had undercut virtually all of the environmental safeguards that had kept Earth's population at a slow but steady rate of growth. Would the population continue to grow or something else happen to curtail the birth rate?

Prior to this time of ever expanding scientific discovery, understanding the environment was more common in daily life. With science able to mold living conditions in unexpected ways, social conscientiousness became more of a focus. The growing sense of responsibility for the underprivileged drove changes in education. Working hand-in-hand with social programs that provided safe places to live, adequate food and medical care, schools became places of equal opportunity for all children regardless of their desire or affinity for learning.

By increasing social opportunities, education had to admit that equal access did not bring about similar outcomes. To level the field, education became about completing tasks not excelling at learning. The most capable students still advanced but information was presented differently and how abstract concepts and higher thinking skills were taught changed. Creativity and innovation were the losers as information became separate concepts without connection to each other and the world.

As time went on, that practice moved into colleges and programs of higher learning. Specialization was preferred to broad-based knowledge that could cross disciplines. College was a coming-of-age experience. Mastering knowledge took second place to achieving an independent lifestyle.

With all the changes that were being made to the environment, was this style of learning going to have what was necessary to unravel the environmental puzzle when the time came? Would there be so much specialization that it became impossible for the new generation of scientists to understand how the environment actually worked? Those concerns would be set aside as another invention would impact environmental understanding. That complication would be called the Internet.

Essential Learning Replaced by Computer Skills

First Apple Computer Model- 1978

The first Apple computer began school in 1978. Initially, it was a great tool to provide practice and motivation. Within a few years, however, learning became more about using a computer than learning from it. Children in lower grades were taught to type and use a word processor as a prerequisite for high school. Whole encyclopedias were just a fingertip away. Who needed to know *how* to make change in the checkout line when the IBM based cash register computed it automatically? Employers had staff with knowledge. They needed employees with computer skills. Basic knowledge of how to use a computer was enough to land a good job.

By the early 1990s, the Internet burst into education. The final argument for rote learning and mastery went out the window as students could simply look up information online. So much of what the human race had learned about nature and the environment was pushed aside to be replaced by socially addictive digital diaries.

If that wasn't enough, the open format of the Internet gave rise to a different kind of ethical issue. False and/or abbreviated information flooded websites. Not only did recent graduates have a different knowledge base than earlier students, but the device they had been taught to trust had become little more than a marketing tool for business.

How Environmental Science Suffered

With the haphazard implementation of the Internet, it was like a century and a half of scientific information disappeared. Books were shunned and libraries were used to gain access to a word processor or social media. Unlike books, which offered consistency once in print, websites had to be maintained and were continuously altered. Computer servers were overloaded with business records and personal blogs. The algorithm-driven popularity search model made it impossible to access dry and strictly factual information--assuming it had been converted to HTML formats.

In the excitement of new devices, users failed to grasp the limitation of computer-driven research. Theories such as greenhouse gases, which had been modified over time by other scientists, suddenly reappeared as gospel truths. Without a solid fact-based education, researchers had no choice but to trust conclusions drawn from computer analysis and online information. As was inevitable, the path that was supposed to bring environmental health, instead led to environmental dangers and poor human health.

Mother Nature's Got This

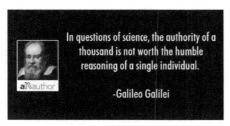

In questions of science, the authority of a thousand is not worth the humble reasoning of a single individual.
-Galileo Galilei

Certainly, admitting that modern science has been compromised is difficult to do. Unintentionally, faith in the computer created a disjointed collection of competing theories. Even with signs such as changing weather patterns and higher incidence of serious disease, users still trust computer modeling over human observation and experience.

When the use of electricity jeopardized the environment in the early 1900s, Mother Nature had the tools to keep the world in balance even if the inhabitants of this planet would not find them comforting. Between 1902 and 1937, no less than twenty-five major volcanic events occurred. During the same time period, approximately 101 earthquakes (6.0 magnitude or greater) and multiple devastating hurricanes occurred. Mother Nature knew full well what was wrong with the planet and used the geological events at her disposal to keep drought and disease at bay as much as possible. A global pandemic took the lives of an estimated one-third of the world's population in order to keep the planet's food supply in line with its animal population.

Today is different. Mother Nature still has the same tools but science has been able to countermand the effects of volcanic eruptions, earthquakes, flooding, wildfires and even a world-wide pandemic. Science still stands by its assessment that the atmosphere is to blame but there has not been any significant abatement of these disasters as air pollution numbers drop.

Correcting climate change will require scientists to back away from their computers and look at a diverse and complicated system of cycles that cannot be reduced to raw data points. Only with collaboration across all scientific disciplines will there be an answer to the question that should have never needed to be asked: ***What is wrong with the planet?***

6

Separating Fact from Observation

Now, the real work begins.

Observation and analysis are not always enough to make an idea factual. A concept must stand up to repeated challenges for it to be considered scientifically accurate. Theories accepted without such trials are a matter of faith not science.

Like medicine or the fine arts, the pursuit of science is a calling. It takes a special type of personality to think logically and abstractly at the same time. But that does not make scientific conclusions automatically correct. Even the most dedicated and educated scientists have their blind spots as well as their strengths.

Scientists today have the responsibility to challenge the work of previous researchers. This is not a matter of disrespect but an awareness that while technology has changed the narrative, it has not replaced human reasoning. Overconfidence in older scientific studies without testing in real situations contributes to inaccurate research in the future. Unfortunately, simply posting theories to the Internet has replaced that most important step in the Scientific Method.

Max Planck and
Albert Einstein

In this chapter, the reader will be asked to consider different explanations for three assumptions which are the foundation of current climate change theory. At present, these theories cannot be physically proven in a real-world setting. Therefore, the ideas presented here offer as valid a position as those commonly believed.

A Thinning Atmosphere

"Old Science" presented the atmosphere as a variable mixture of gases, clouds and particles that separated into layers. Currently, the Earth's air mass is viewed as a stripped down version that is more delicate than previously thought. Here, for the sake of argument, both versions will be considered.

With physical testing limited by the vast and unreachable nature of the atmosphere (estimated as being as much as 300 miles above the surface), both theories agree in only one way. Known to be densest near the Earth's surface and thinner as it radiates out into space, this characteristic can be proven through a variety of measures. The difference between the two concepts comes from the creative and out-of-the-box studies which have drawn some interesting conclusions as to why and how this happens.

One study, with its results posted to a NASA.gov web page in December of 2003, fanned the flames of climate change concern. Detailing studies from the Naval Academy Observatory and space photography from a variety of returning and orbiting spacecraft, the article drew a perhaps *accurately wrong* conclusion from the spellbinding visual images. Observing that there had been a 10 percent decline in the density of the upper atmosphere over the previous thirty five years, the study concluded greenhouse gases were a contributing factor. With no explanation as to how carbon dioxide could cause such an event or what may have happened to the missing gases, the posting has become the basis for other equally unproven claims.

Atmosphere Photo from NASA article - December 2003

At the time of this writing, the NASA article is listed as being archived and no longer updated. It also quantifies the science by labeling it as "based on the best science at the time."

What exactly does that mean? Does it imply the conclusion was incorrect or was the study's design faulty? While still posted and available, the article, intentionally or not, supports the concept of a thinning atmosphere but does not provide an experiment-based cause for this situation. Rather than speculate, science should consider what is known about the lower atmosphere and the role that carbon dioxide plays in it.

In keeping with known facts, carbon dioxide is produced by animals and used by plants. Unlike oxygen which can be in different forms, carbon dioxide is a three-atom molecule with an atomic weight (44.09 g/mol) that is greater than most other gases. Its weight alone makes it fall to the earth and stay where it is most needed—where plants live.

In turn, carbon dioxide that finds itself farther up in the atmosphere during air travel or through volcanic eruptions would predictably fall to Earth. In the NASA photos, this line of carbon dioxide is very clearly visible near the cloud layer.

A similar line, called the tree line, marks the upper most part of a mountain range where complex vegetation stops growing. The tree line has been known for years to be the altitude at which the atmosphere becomes too thin to support plant life. Since carbon dioxide is needed for plants to survive, it can be inferred that the tree line and level of carbon dioxide in the atmosphere are similar.

Given that carbon dioxide has always been a dense gas and its location in the atmosphere has always been limited to an altitude similar to that of the tree line (not to be confused with tundra which is based on latitude), it seems quite the leap to blame carbon dioxide for a less dense atmosphere.

Could there be a more subtle cause that might be contributing to what is described as a "thinning" atmosphere? Discussed further in Chapters 7 and 8, the use of lighter-than-air gases in industry may have more to do with this issue than the role of carbon dioxide. For now, the theory of greenhouse gases needs to be scrutinized.

Greenhouse Gas Theory

By now, this book has made its position on greenhouse gas theory clear—it is Internet fiction. In more descriptive terms, like a thunderstorm on Mt. Olympus that became the stories of Ancient Mythology, the Internet took one bit of scientific knowledge, chosen to fit a particular void in environmental science and turned it into a cult following that neither proves global warming nor solves it.

Now it is time to justify that characterization with something more than conjecture.

Greenhouse gas theory is credited to an English/Irish scientist named John Tyndall whose major focus was solar radiation, physics and geology. Regardless of where information about greenhouse gas theory is found, it is described in short general phrases and includes one quotation that seems more observation than scientific finding. Despite this lack of evidence, "greenhouse gases" are believed to be the cause of everything from melting glaciers and rising sea levels to heat waves and volcanic eruptions.

How does one disprove something the Internet says is real but for which there is no proof?

Before looking at Tyndall's theory, it might help to understand where the term *greenhouse gas* came from. What is now commonplace in 2021, the greenhouse had only been in existence for about 20 years before Tyndall named his study.

The invention of a wealthy French botanist, Charles Lucien Bonaparte, the "glasshouse" was constructed of glass and iron for the purpose of year-round propagation of medicinal herbs. Many are familiar with the saying, *People who live in glass houses shouldn't throw stones*. That phrase may have begun during this time period. Over time, the term became "green" house because of the green plants grown inside.

Regardless of where its name came from, the "greenhouse" quickly became a fascination and sign of wealth for many families in Europe.

What brought the "greenhouse" to the global stage was London's Great Exhibition of 1851. The **Crystal Palace**, the largest such structure of its kind at that time, was built of cast iron grid work and small panes of glass. For six months, the event welcomed 14,000 exhibitors in a 990,000 sq. ft. space. Scientists the world over were taken with the building, particularly those who studied solar radiation. Tyndall was one of many to study the phenomenon.

Great Exhibition of 1851- Crystal Palace Interior

Also of interest might be the working conditions of a scientist in the 1850s. A legitimate vocation, and with research becoming an increasingly important part of a professor's duties, Tyndall was a sought after lecturer and author. His expertise in multiple subjects made him a valuable member of a faculty or as a guest lecturer. As competition increased in the academic world, scientists gained fame and recognition by making discoveries. The first to write a paper or publish a journal was praised even if the conclusions were later altered or found to be inaccurate.

Further investigation finds that Tyndall's proximity to London would have made it easy for him to visit the **Crystal Palace** and experience the warmth of a large glass building fueled by nothing more than the sun and the air in it. Use of the term "greenhouse" would have drawn attention to his work with radiation and gases while giving it an up-to-date and trendy feel. Even then, professors were bound to make their programs look good as well as be good.

As his work is considered, Tyndall's accomplishments are significant. But do they actually prove what the world believes they do?

There is no doubt that the scientist measured the amount of radiation each of three gases (carbon dioxide, water vapor and ozone) absorbed. He also measured the amount of time it took for each to cool down. All of that is recorded in his notes and journals.

But, these experiments were at the beginning of the study of thermal radiation. Like so many scientists of the day, Tyndall worked in his own community, probably alone, while other scientists were making the same discoveries in locations around the world. Each worked without the ability to quickly converse with each other. Because of growing competition between programs of higher learning, it may not have been advisable to collaborate with other scientists but better to adapt one's findings after reading conflicting theories in professional journals.

What appears different here, and what might have attracted climate change advocates to this theory, is Tyndall's exclusive focus on gases along with his hobby of high altitude climbing. In today's context of specialization, Tyndall's theories would appear connected, when in the context of his life, they might not have been. Taken out of context and checking off multiple boxes for the climate change frenzy, Tyndall was the perfect scientist to use to justify an otherwise unproven theory.

Here is why Tyndall's theories match some ecological conditions but not all.

Ireland and England have maritime climates. By definition this climate type is more moderate in temperature and does not have well-defined seasons. Weather during Tyndall's lifetime would have been similar to that of Washington State. For him, the occurrence of frost would have been rare at ground level but common on his hikes into higher elevations. His apparent conclusion that carbon dioxide and other gases used in his experiments were responsible for the absence of frost was based on accurate observation and wrong interpretation. Environmental activists in coastal climate, who may have never seen annual frost, would have easily been convinced by the similarities between their situations and Tyndall's.

To put energy use into context as well, the use of electricity would not reach the British Isles until twenty five years after he published his findings and just ten years before his death (1893). The only way for Tyndall to heat his home or cook foods was through the use of fossil fuels (coal, oil, peat, gas or wood). These fuels would have released a steady and warm mixture of smoke into the atmosphere. It would have only been when Tyndall climbed above the cloud layer/ tree line that he would have experienced frost. Carbon dioxide alone did not keep the surface from freezing. A sea level climate and use of fossil fuel were responsible for the warmth Tyndall attributed to radiation.

By 1900, Tyndall's findings would be replaced with those of thermal radiation. Acknowledging that radiation turns into heat, thermal radiation does not limit that ability to gases alone. In that way, the greenhouse effect attributes the warmth of a greenhouse to radiation turning to heat when it passes through the solid glass, not just heating the air inside. If

Tyndall had used bell jars to capture his gases, as many scientists did in those early days of gas study, the glass jar would have replicated the glass panes in the greenhouse and further confused Tyndall's findings.

How greenhouse gas theory came to be accepted can be explained in three ways. First there was a government which legislated policy as if science was something it could control. Then there was the public perception that scientists could not be wrong, a result of a decade or more of stunning accomplishments such as the first open heart transplant and walking on the moon. Last on the list was the new, unregulated and unsupervised information system called the Internet which alongside the computer changed how people gained information about the world around them. With just enough knowledge to be dangerous, the Internet sent the world into a maze of half-truths that has now become 25 years of incorrect assumptions about the role of air pollution in the environment. Overconfidence in itself resulted in an educated, technologically advanced society which targeted a small group of always present, sometimes smelly gases as the cause of climate change?

A look at a hard copy of Grolier's Encyclopedia printed in 1991, the year the Internet started, provides some insight. Neither the name, John Tyndall, nor the term, greenhouse gas, is found in the index. The volume does contain information on what it calls a "popular term" used to describe the belief that lower level carbon based gases and water vapor keep the earth from freezing. While explaining the basis of solar energy being absorbed by the earth, there is a tone of caution when the *Greenhouse Effect* is tied to the belief that these gases are what keep the planet warm. Not going so far as to discredit the assumption, the reference points out the rising support for the unproven position and the lack of information about the atmosphere's carbon cycle. Written twenty one years after clean air legislation was passed, the passage mentions that no studies on the individual effect of these carbon based gases had been done.

A concern whenever older scientific information is interpreted without considering the context, this conclusion that carbon dioxide is the cause of global warming is accurately wrong. Accurate findings from one man's work were wrongly used to explain and confuse people from the real cause of a changing climate. Now faced with a world that believes in a philosophy that could negatively impact every living creature on the Earth, science must decide whether to admit its mistakes and encourage changes in legislation or watch as people support a dangerous and faulty path.

Oxygen and Ozone

Not to be left out, the ozone layer receives much attention as part of climate change discussions. What is ozone? Why is it an EPA restricted emission and why is it so important to human life at the same time?

As an element, oxygen can be quite social. It helps to think of oxygen as a teenager. The element does not like being alone. It prefers a group setting and it is not always picky as to which elements it joins. If no other partners are available, oxygen joins up with itself to make a two atom molecule (O_2). This form of oxygen is what most think of as breathable oxygen. A three atom molecule of oxygen forms during chemical processes that release large amounts of oxygen atoms. This three atom molecule is known as ozone (O_3) and makes up the ozone layer.

Where oxygen is found plays a big part in how its role is interpreted.

Situated at least 9 miles (15km) above the Earth, the ozone layer is where most of Earth's oxygen congregates. In what might be a connection to Tyndall's radiation experiments, ozone is believed to absorb radiation to keep it from traveling on to the planet's surface. Some differences exist about whether absorbing such radiation protects the Earth or increase surface temperature. Regardless of whether ozone absorbs or reflects radiation, this collection of ozone molecules may have a much more practical and essential purpose than as a radiatation shield.

Oxygen has many faces and personalities. It makes up 65% of the human body and joins with hydrogen to form water. It is a primary component of food and medicines. Too little available oxygen causes brain damage and suffocation. Too much oxygen at one time causes lung damage and impacts the central nervous system. Oxygen also has a major role in how the environment works.

In Nature, high levels of free oxygen make everything faster. Fire burns hotter and is harder to manage. Metals rust quicker. Engines have more power and muscles work harder. Oxygen has a good and bad side. This may be the environmental reason why the ozone layer is so high up in the atmosphere.

How ozone and oxygen react in the atmosphere may not be as predictable as chemists would expect. Like all chemical reactions, ozone is part of a cycle that ensures adequate oxygen for people and other animals.

Looking at this from an environmental standpoint instead of one based on greenhouse gas theory, what is perceived as a dire situation may be the environment doing what it was meant to do. In areas of heavy plant growth, the levels of ozone are, and should be, higher – that is why wildfires burn so fast and hot. Plants use photosynthesis to break down carbon dioxide and water which are converted into sugar and starch. The process results in extra oxygen being released by the plant. For efficiency, plants sometimes release oxygen as ozone rather than one atom at a time.

By restricting ozone production, the EPA disrupted the ecosystem's fail-safe method of providing adequate oxygen for all life. Plants are not distributed equally across the globe. Neither are people. Without production of ozone from fossil fuel use, every breathe taken in a crowded location is a competition for life. This may be where the ozone layer, and its depletion, has come into play to provide for those who choose to live in unsustainable locations.

One conclusion for the depletion of the ozone layer may be a case of 'just enough knowledge to be dangerous." The discovery of fluorocarbons in the 1940s and 1950s revolutionized the spray can industry. As science recorded a decrease in atmospheric ozone, several cultural events occurred simultaneously.

One of these events was the use of hairspray which skyrocketed due to the popularity of the beehive hairstyle (1960). Keeping the style in place required layers of the lacquer-like spray, particularly if hair was shorter at the start.

Coincidentally, good health care reduced death rates which increased life expectancy in most countries. All at once, the world's population took a tremendous growth spurt. In 1960, the world's population was estimated at three billion people. Fourteen years later, it had risen to four billion and by 1984 when non-essential use of fluorocarbons was banned, the population was estimated to be nearly five billion people. Today, the planet's population is estimated at eight billion people.

Can the argument really be made that fluorocarbons damaged the ozone layer or is it much more likely that human intervention caused this problem? Not only did the environment need to triple its supply of available oxygen for a growing world population but the agency tasked with protecting the environment limited naturally occurring ozone emissions. In a very real sense, fluorocarbons may have had nothing to do with the shrinking ozone layer.

Until science and government look at how the environment functions rather than what explains human perception, the environment will always be in a position of using massive and life-threatening events such as volcanic eruptions, earthquakes and widespread drought to maintain balance. With more scientists now calling for this to be labeled a "Climate

Emergency," what type of rush to judgement will science create to explain frequent volcanic eruptions and massive drought?

7

Another Look at Gases

Sometimes the best way to explain a scientific principle is to compare it to something utterly ridiculous. For that reason, imagine the atmosphere as an Everlasting Gobstopper.

Interior Everlasting Gobstopper - Breaker Confections, Chicago, 1976

In the children's novel, *Charlie and the Chocolate Factory*, the Everlasting Gobstopper is a mouth-sized jawbreaker with multiple but distinctly unique layers. Like the atmosphere which has been a steady part of the Earth since the beginning of time, the fictional version of the candy is said to never disappear. As seen in the image below, the real version of the candy helps to show the diverse nature of an environmental feature that is all but invisible. On the outside of both the candy and the atmosphere, a thin but resistant layer acts as a protective barrier. In terms of the Earth, this barrier keeps the atmosphere from dissipating into Space. Current textbook-style descriptions divide the atmosphere into layers based on their perceived function. In contrast, the gobstopper example has dozens of thin, well defined rings which correspond to the many gases that make up this air mass. Rings near the outer most edge represent a group of fourteen substances called lifting gases which will receive some attention in short order.

The larger ring near the middle of the sphere represents the Ozone layer, a natural storage unit for unused oxygen. The more central rings become thicker. It is here that gases which are the focus of air pollution and climate change discussions occur.

In keeping with its chemical properties, each gas or vapor rises only as far as its atomic weight will allow. Wind currents and the Earth's rotation continuously stir the gases to make a blend that is breathable as well as functional.

Realistically, atmospheric gases and their locations above the surface fluctuate in amount but have always been a part of the atmosphere's makeup. Focused so heavily on carbon emissions, those who believe in greenhouse gas theory are unintentionally discounting the importance of the many gases which comprise breathable air.

A Potential for False Narratives

For many years, a sense of duplicity has hung over the topic of air pollution and its relationship to illness. As the asbestos and tobacco industries went about covering up the link between their products and some types of cancer, the EPA was supporting research which gave credence to its position that air pollution caused thousands of deaths each year. Was this a move to justify restrictions or to protect two large industries from being held accountable--as they knowingly put lives at risk?

The National Center for Biotechnology Information under the umbrella of the National Institute of Health (nih.gov), has catalogued countless air pollution-related reports over the last four decades. As can be typical of computer generated data, many studies involve small tightly defined groups of respondents; data-collection methods that are open-ended or easily swayed; and a narrow focus that would almost guarantee a positive correlation to the targeted outcome. In turn, the EPA took this information, extrapolated the data into massive numbers of deaths and declared the findings a valid reason for combating pollution. These figures are still posted on the EPA's website even though the agency has claimed success in ridding the atmosphere of at least half of the pollutants as of recent figures. Would a survey of actual death certificates confirm or refute these claims?

As a government agency which answered to the President and one that was able to determine its own success, its staff had the flexibility to interpret such data in any way needed to meet the agency's mandate. With fear and data based on questionable sources as its primary tool, the EPA was able to convince the public that its decisions were based on their well-being, not a political agenda that changed with every incoming president.

How much longer will the American public accept the word of government without taking a good hard look at its validity? Will the current Congress blindly support a massive investment into unproven environmental theory for only one reason, to win votes? The time for that answer may be coming sooner than imagined.

Lifting Gases and Their Important Role

Fourteen elements or compounds fit the definition of a "lifting gas." Considered to be lighter-than-air and capable of lifting dust and water vapor into the upper atmosphere, these gases perform a very important environmental function.

Upper Atmosphere Weather Collection Balloon

Hydrogen is well known to most people. Others like acetylene, ammonia, and fluoride are also common but only as products used in daily life. At least two of the fourteen, methane and carbon monoxide, have been a focus of clean air legislation in the last 20 years, and three others, nitrogen, neon, and helium, create millions of dollars in annual revenue when used in cryogenics and manufacturing. It is the use of these gases for industrial and healthcare use that may have contributed to the 10% drop in atmospheric density previously reported by NASA. As non-lifting gases such as oxygen and carbon dioxide are removed for similar purposes, it becomes difficult to say how much of the atmosphere has actually been mined for industrial purposes.

One important environmental role these gases play is in the formation of radiation shielding clouds. Explained in detail in Chapter 8, lifting gases partner with small particles and water vapor to reach higher altitudes where the water vapor freezes. Rather than producing weather, this type of cloud formation produces a wispy-looking network of ice crystals which, like glass, have the ability to reflect radiation and protect the polar ice caps.

Not covered by the EPA's mandate, US imports of semiconductors and electronics, which rely on rare gases as part of the manufacturing process, happen not only for cost reasons but to escape regulations. In contrast, according to WorldBank figures, the US exported $7 million dollars worth of medical oxygen in 2019 which is likely a small fraction of what the country used that year. With the technology available to extract oxygen from the atmosphere on demand, it is difficult to know how much oxygen is removed and how it impacts those who must breath the less concentrated mixture. Removing gases from the atmosphere, even now, may not be considered harmful by science. Climate change advocates have already designed ways to remove carbon from the atmosphere taking full advantage of the fear that carbon dioxide will lead to human extinction. Such economic answers to environmental problems only serve to further weaken that part of the Earth that protects all of life.

With the EPA's focus on preventing only the environmental damage it can see, a disconnect exists between policy and real world environmental application. As the EPA approaches zero tolerance in eradicating functional gases from the atmosphere, the danger is very real that these decisions will continue to compromise the planet and public health.

Bottled gases ready for market

In addition, air rights are different than mineral rights. Changing the quality of the air overhead only qualifies as pollution if something unhealthy is added to it. Taking essential gases out of the atmosphere is not considered pollution. From the creation of dry ice (frozen carbon dioxide) to filling tanks of nitrogen, hydrogen, acetylene, and argon for metalwork, extracting gases and bottling them is big business but one that has environmental consequences.

Pollution: Urban vs. Rural Considerations

Environmentally, the ***City Mouse / Country Mouse*** fable may have a good lesson for environmental advocates. For those unfamiliar with the tale, two mouse cousins, one living in the country and the other living in the city, decide to visit each other. Both find the other's way of life nice in some aspects but disconcerting in others. While the story ends with both feeling grateful for what they have instead of what they don't, would the outcome have changed if the tale had transpired over centuries instead of days?

For more than a century, urban areas have taken more from the environment than they have returned. Feeling monetary compensation that barely covers the cost of production is a fair trade for food, water, and raw materials to build and use, cities fail to recognize they can not survive without the support of rural ecosystems. In terms of the environment, nice restaurants, museums, theaters and shopping have no value. Yet, unsustainable urban populations are the voices of climate change advocacy and policy.

Proclaiming concern for the environment with legislation and restrictions, the image of environmental responsibility has become a part of the urban brand, not its practice. In return, rural communities must follow rules which are largely unnecessary and absorb costs that support unsustainable lifestyles and practices.

Rather than understand that rural communities are fed up with preferential treatment that robs their communities of resources and tax dollars, politicians, social justice advocates and the media fabricate illusions of racism that again deflect environmental responsibility away from urban users and piles it onto those who provide them with food and water.

The line in the sand is clear. Environmental law and policy identify select gases as pollutants. These decisions were made almost exclusively to placate urban areas and public perceptions. Cities want to be bigger but they do not want the trouble of producing their own food and water, maintaining plant life which cleans the air and cleaning up the waste they create. Cities use government legislation to make their problem everyone's problem and rural areas that ultimately care for the environment suffer as a result.

Of the five specific substances which are targeted by EPA guidelines, all occur naturally and not solely because of gasoline engines or industrial operations. Here are the benefits of each gas and how they may be less of a pollutant and more an essential environmental ingredient that will help end climate change. Charts showing the drastic reduction in these gases can be found in the Informational Graphics section in the back of the book.

- Carbon monoxide is a lifting gas that adds density to the atmosphere. A denser upper atmosphere may protect the planet from radiation. With the focus of the last twenty years being on reducing carbon emissions, research has failed to look into how carbon monoxide responds naturally in the atmosphere. It is possible that, like lightning producing nitrogen dioxide/oxide, conditions exist in the atmosphere that promote bonding with free oxygen and to produce carbon dioxide.
- Nitrogen oxide compounds are produced by burning organic fuels as well as lightning strikes. Farmers have long held that nitrogen oxide compounds in the atmosphere transfer nitrogen to the soil when it rains. Nitrogen is a necessary nutrient for nearly all plants and is required for all life to make protein. Looking at its impact on human life, nitrogen oxide compounds may have a direct impact on mental health. As air-borne nitrogen oxide compounds have dropped, the incidence of mental illness and violence have increased globally. Did people feel happier fifty years ago because the culture was fundamentally different or because the same gas that is used to relieve anxiety in dental offices was more prevalent in the atmosphere?
- As discussed previously, ozone is a controlled substance and one that, at the same time, causes concern about its depletion in the atmosphere. To recap, while medical personnel see ozone as dangerous, other sciences understand that ozone falls apart or "decays" in the presence of water. In other words, like the other gases restricted by the EPA, rain causes ozone to become free oxygen which can then be used safely by animals The more that the envi-

ronment is examined as a whole, the stronger the evidence becomes of its highly interactive nature.
- Sulfur-based gases include those that combine sulfur and hydrogen as well as those that pair sulfur and oxygen. The EPA only tracks sulfur dioxide. Like nitrogen, sulfur is an essential nutrient for plants and animals. Its atmospheric reduction has been linked to lower crop yields and some health conditions. Interestingly, hydrogen sulfide, commonly known as sewer gas and monitored by the Occupational Safety and Health Administration (OSHA), is now being considered as a cheap and effective treatment to lower blood pressure. With its findings reported in a 2015 article published on the PubMed section of the National Institute of Health website, the study is one more example of the inconsistency of federal clean air policy. As for how ambient sulfur factors into wildlife and natural habitats, gaseous sulfur would help explain how herbivores (those that eat only plants) can manufacture proteins that support life. Low birth rates for endangered species may have more to do with a low sulfur atmosphere then reduction of habitat. As the World fears for an upcoming mass extinction of species, it seems more helpful to return sulfur to the environment rather than calculate numbers of deaths. Computer analysis is not a crystal ball of things to come. Like Scrooge, humans can change their ways and produce a different outcome.
- Particulate matter, otherwise called dust, is tracked by the EPA. Ironically, this dust is frequently made up of carbon and heavy metals that can be used by plants from the soil. Particulate matter also takes on a critical role in how clouds are formed and how well they disperse water around the planet. This cycle will be the focus of Chapter 8.

A Few Words About Methane

A carbon-based gas that has been tied to the climate change discussion, methane production has been suggested as a reason to adopt a vegetarian lifestyle in the hope of reducing greenhouse emissions. That said, not only the by-product of commercial farming, methane is produced when organic matter from plant, wildlife and human sources decompose as well. Regardless of source, farmers have been subjected to costly regulations in the effort to control methane and other contaminants. What actually happens when methane enters the atmosphere is rarely considered.

Like ozone, science has postulated that methane changes form over time. Some of these changes may occur in the air closest to Earth. More importantly, it is one of the fourteen gases which has the ability to rise above the cloud layer and provide density to the upper atmosphere. There it may remain for many years or react with other gases as part of the environment's cyclical nature. Also used in industrial applications to fuel engines and boilers, methane seems an odd material to environmentally blacklist.

The Concept of Air Quality

A discussion of air pollution would not be complete without touching on how air quality is measured. An extension of the EPA's guidelines and monitored by NOAA (National Oceanic and Atmospheric Administration), the tool is known as the Air Quality Index or AQI. Targeting the five regulated air components mentioned above, it measures the amount of each substance and, using some rarely documented formula, transforms the numbers into a numerical score that alerts the public to potential dangers. Pollen and humidity can sometimes factor into the score but it is largely a representation of the EPA's goal to reduces what it calls *pollution*.

NOAA has been collecting weather data for decades. Small weather stations are located across the country that monitor temperature, wind speed and barometric pressure. Often located in federal or municipal locations such as post offices and airports, NOAA has access to data that appears to represent current weather fairly. But the measurement of air quality may not be measured or calculated with the same level of detail.

The EPA, in its reporting of average levels for specific pollutants, lists as few as 21 and a maximum of 192 locations used for collecting data. The equipment needed to measure levels of specific compounds can be very expensive and air samples can vary widely from hour to hour depending on the proximity to a pollution source. However, many weather apps readily list Air Quality measures without specifying location. Some websites such as IQAir.com post air quality numbers in advance. Assumed to be computer-generated based on weather trends and historic data rather than actual measurements, the AQI might be more of a publicity tool to validate the EPA's role than an accurate measurement device that protects the health of citizens.

In keeping with this book's judgement that science has been compromised by computer analysis and economic priorities which were substituted for actual observation, the Air Quality Index should not be considered a reliable measure of the health of the atmosphere. Until the government accepts that its decisions are a factor in current climate change, the world's population will never enjoy the health and well being that it once did.

8

A Discussion of Clouds

Unintended consequences are not always the result of unintentioned actions. As ridiculous as it sounds, what is considered climate change could be the very real consequence of intentional decisions that were based on how the environment looked rather than how it functioned. Could it be that the answer is in plain view? Could the answer be something as simple as a "cloud"?

Clouds and Their Function

Besides being pretty to look at, clouds have more than one environmental purpose. Their most obvious role is as part of the Water Cycle, a scientifically recognized continuous process that redistributes water from one place on the planet to another. A second, and seemingly forgotten function, is that clouds also protect the Earth from large amounts of radiation which becomes heat.

Surely, it cannot be as simple as something happening to clouds that has made the weather erratic? No, this problem is more than the absence of certain clouds in the sky. However, any climate change solution that attempts to solve the problem without including clouds in its formula is sure to fail.

The Water Cycle

The Water Cycle starts with the compound known as dihydrogen monoxide, otherwise known as *water*. There is something unique that happens when two atoms of hydrogen and one atom of oxygen combine. Separately, each of these elements has a dark side but together they make a partnership that gives life to this planet and provides flexibility within a strong purpose.

Taught as part of a science curriculum in the early grades, this essential cycle seems simple enough and yet, the world deals with devastating drought and flooding rains, the two extremes of a malfunctioning water cycle. Could it be that what was called cyclical fluctuations have been indications of climate irregularities throughout the ages? In its simplest form, the Water Cycle has four stages:

1. Water evaporates
2. Clouds form through condensation
3. Rain falls from clouds
4. Rain is used by plants/animals with run-off flowing to the ocean. Repeat.

Doing more than redistributing water, the water cycle fulfills multiple functions that seem invisible to anyone but the most avid admirer.

How Clouds Form

Clouds form when water vapor condenses on some type of airborne particle. Condensation means that water molecules stick to a cooler surface—like the wet film on the outside of a cold drink glass. Dust and other particles in the air act like the glass. Without dust particles, water vapor hangs in the air until the air temperature causes it to change forms, either as precipitation or as dew. This example may help demonstrate the need for dust particles in cloud formation.

Think of humidity as children riding in a minivan or bus. One or two children on the bus are no problem but as the bus fills with passengers, the space between them diminishes. The driver hopes to get to the destination before someone cries out "He's touching me," Like the bus that then erupts into chaos, it only takes one slight bump between molecules of water vapor to cause an afternoon shower. This is why dust particles are so important.

Like children who have a distraction—a snack, doll, video game or headphones—water vapor is more stable when it holds on to a particle of dust or pollen. When water vapor and dust partner with a lifting gas, the result is a stable cloud high above the ground.

There is much to know about clouds but here the discussion will be limited to where and how clouds form. Weather producing clouds are typically found near Earth's surface while radiation shielding clouds exist mainly over the poles and above those that produce weather. With clouds so difficult to reach and study, some information about them that is presented here is more observation than proof.

Clouds Connection to Climate Change

As electricity replaced fossil fuels, users would have been unaware of the connection between smoke and a healthy environment. After all, meteorology and much of what is known about gases had yet to be discovered. The unintended consequence of these two events was to create a disruption in the water cycle that would build up into a major period of drought for the United States.

California is a prime example of how "clean air" works against the environment. The state has been a leader in clean air policy for decades. It is the years when air pollution was at its greatest that provides the best example. With moist, ocean air in unlimited supply, particles and gases from air pollution formed stable clouds. These clouds carried large amounts of water over the Sierra Nevadas where those clouds would deposit their moisture as rain and snow. This snow pack would then supply multiple states with spring runoff.

As the state adopted stricter and stricter rules on fossil fuel use, fewer dust particles and less air pollution were available to grab the moist ocean air and redistribute it over the mountains. Drought is a natural consequence of a world of limitless sunshine. With every energy saving, clean-air-preserving measure, California--and the country--comes closer and closer to repeating the Dust Bowl droughts of the 1930s. The following is a pictorial representation of how air pollution benefits the Water Cycle. Part of a paper published in October 2019 by a collaboration of Indian scientists, ***Aerosol-orography-precipitation – A critical assessment*** details how and why air pollution creates a more productive water cycle.

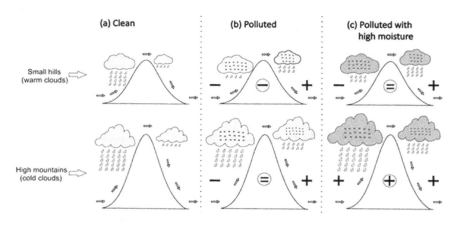

Illustration of upslope precipitation w/wo pollution
Science Direct website Aerosol-orography-precipitation – A critical assessment

Nature's Dust Mop

Long before the concept of climate change became such an emergency, clouds were a gauge of the upcoming weather. Together with the wind, they aided in circulating dust, smoke, and fumes along with moisture. Before anyone knew how the movement of clouds helped the environment, everyone knew that without them air became stale and contaminated with sickness. Like the quick swipe of a dust mop, the wind and the clouds moved the air around until it was fresh and clean. Like a damp mop, a good rain would clear the air of all things unsavory.

Nature's Parasol

A high-altitude version, noctilucent clouds exist miles up in the atmosphere and above weather making currents. Little is known about them except they are named for their tendency to change colors before sunset. Believed to form over the polar ice caps and drift to more temperate latitudes, these clouds act as a lady's parasol by protecting the Earth's delicate, colder regions. Their light reacting construction, possibly as ice, reflects radiation back into space before it gets a chance to be absorbed by the planet's surface. To review, radiation that is reflected back into space does not warm the planet. Only when radiation connects to a solid surface is heat produced according to thermal radiation theory.

Not coincidentally, noctilucent clouds share their high altitude home with lighter-than-air gases. As has been discussed in a previous section, diminishing supplies of such gases in the atmosphere make it more difficult for these clouds to form. Could this be but one more wound to an endangered atmosphere? Which does the world want more, a healthy environment or the computer-chipped devices which can only be produced by weakening the atmosphere?

Forming in much the same way as weather producing clouds, these translucent varieties offer more than protection from solar radiation. As a sign that the environment is healthy and happy, these clouds produce an evening light show as they refract the sun's last ray's and create beautiful skyscapes at the end of each day.

What This Means for Climate Change

Upper altitude clouds in sunset sky

Never realizing that its own decisions and policies created climate change, government now seems focused on fixing it. Based on a strong belief in carbon footprints, greenhouse gases and zero tolerance for carbon-based emissions, how likely is it that future legislation will produce a successful outcome?

Before the EPA's campaign for pure air and water, impurities would have been a part of daily life. In every place and every time period, wherever fuel was burned, dissolved minerals and gaseous compounds would have been a part of usable air and water. That history alone gives a solid reason to look at something other than air pollution as a cause of climate change.

Too often, the average individual believes the elementary school version of important scientific processes. In reality, these cycles are complicated, fine tuned, automatic partnerships within a well-designed environment. Was fossil fuel to blame for air pollution or was it some other factor that caused the environment to stop working? Before advocates point fingers at a convenient scapegoat, it might be worth looking at what this rush to judgement has cost humans in terms of draining healthcare costs and declining personal health.

The Real Risks to the Environment

Low level water vapor in the form of mist, fog and haze might be part of the Water Cycle but it is not the most efficient way to redistribute moisture and nutrients. Until clouds form at a high enough altitude to draw moisture over mountains and long distances, dry areas stay dry and damp areas stay damp.'

Does it matter what clouds look like as long as it rains? That may be the very question that needs to be answered. Consider the following which may help clarify the situation.

* Without clouds, more radiation hits the earth and raises surface temperatures.

* Without high altitude clouds, water vapor produces shorter, more frequent periods of precipitation. Such moisture tends to fall quickly and does not always soak into the ground.

* For young plants, air pollution provides gaseous nutrients that might be missing in the soil. As they mature, plants use the same gases to build protein that becomes food for animals.

* Without clouds, water vapor stays closer to Earth and makes it harder to breathe.

* Limiting the production of gases without understanding their purpose may alter the composition of the atmosphere and make it more difficult to breathe.

* Without clouds, millions of gallons of water are returned to the ocean instead of being stored in the atmosphere where it acts as a shield against radiation. Are sea levels rising because of melting ice caps or because clouds cannot form to store that water in the atmosphere?

* Gaseous nitrogen and sulfur, distributed by clouds, may help herbivores (species that eat nothing but plants) maintain adequate protein levels.

* Volcanic eruptions, mysterious gas plumes and other geological events like wildfires are to be expected as the environment attempts to repair the damage that has been caused by EPA regulations.

What about Today

Fighting the natural order of the world has historically been a losing battle. During the early days of the pandemic, the fortieth anniversary of the eruption of Mount St. Helens in Washington State went almost unnoticed. May 18, 1980 was an eye-opening moment for most US citizens.

Used to earthquakes in Southern California, annual hurricanes in the Gulf and Atlantic and a handful of well-behaved volcanic tourist attractions in Hawaii, American and Canadian residents were unprepared for the force of the explosion and magnitude of the devastation that followed two months of minor geological activity. As scientists look back at the event, their focus is on what they missed, how much they could have learned if they had the equipment available today and how better prepared they are for the next one. (www.usgs.gov) After multiple, unexpected and devastating eruptions around the globe in the first few months of 2021, it is hard to validate the conclusion that eruptions can be predicted and lives and property saved.

Mount St. Helens Plume (unknown date)

Documentation of that day and the weeks after covered the estimated height of the plume, the number of acres covered by as much as three feet of ash. and the rate at which the ash cloud traveled across the globe. How the eruption change the pattern of the weather was not much of a concern after the first few weeks.

Was 1980 a record breaking weather year because of the eruption of Mount St. Helens or was it the result of ten years of environmental policy that was so drastic that EPA Region 9 (Southwest USA) trends show as much as a 75% drop in sulfur dioxide levels between 1970 and 1980.

(See graph in Informational Graphics Section). If similar reductions of air pollution were documented across all criteria, it would have been difficult for the environment to form any type of sustainable cloud cover on a local level.

Urban areas that were first to adopt electricity developed land quickly without concern for environmental limitations. Still holding tight to the belief that drought and poor public health are due to something in the air, climate change alarmists are at least half right in that conclusion. Drought, poor health and increasing levels of high ultraviolet radiations are results of changes in the atmosphere, just not the way they assume. As explained here, it is not what is *in* the air that is the problem. It is what is missing that has caused these outcomes.

Unfortunately, the most devastating part of this situation is not that urban residents must live with these outcomes but that rural areas which had little or nothing to do with such decisions have been forced to share the human toll of devastating weather events and declining health--all because cities had the legislative power to push their agenda as being equally beneficial for all. Environmentally, there is no equal when comparing rural and urban communities. How long will the City Mouse be able to convince the Country Mouse that urban lifestyles are somehow better for all when they are environmentally unsustainable? When will the federal government recognize that it is the urban lifestyle that endangers the planet, not fossil fuel use?

9

Air Pollution Provides Sulfur

Understanding *how* clouds and the Water Cycle work together is not enough to understand how climate change happened. In a good detective novel, the plot winds its way through explanations until the main character finds "the smoking gun." That is what this book has tried to do. It is time for the last tidbit of evidence to be unveiled.

So ubiquitous in the ecosystem that science took it for granted, sulfur provides the secret ingredient that makes the environment healthy for all living things. This illustration may help in understanding this elements importance to life.

Periodic Table Symbol for Sulfur

The Importance of Sulfur: The Domino Principle

Imagine the world as a large gymnasium floor. Across the floor are thousands of black dominoes in intricate designs. Spaced to fall like a grand domino exhibition, each pattern represents a type of life form. Throughout the designs, single yellow tiles (sulfur) are placed at important junctions. With a touch, sulfur facilitates the creation of life. Without it, the patterns break apart and cease to respond. In reality, nature has backup plans for when sulfur is not readily available but those alternatives often come with limitations and side effects that impact how well life is lived.

What is Sulfur?

One of the first elements identified, sulfur is easily spotted by its light yellow color. It has an atomic number of 16 and a symbol of S. Like oxygen, sulfur can be found in compounds that are solids, liquids and gases. As a gas, it often has that *rotten egg* smell. In other forms it serves many environmental purposes.

Nearly a hundred years after the *Titanic* sank, a higher than recommended concentration of sulfur was found in the ship's steel, finally explaining why the hull ruptured instead of denting on that night in April 1912. As in the production of steel, a little sulfur goes a long way and impacts many life functions. Too much sulfur may cause harm but too little sulfur can be equally damaging. The keyword for the environment is *moderation* which can be hard for modern civilizations to embrace.

Sulfur dioxide was labeled as a pollutant by the EPA soon after the agency's formation. Without considering that the smelly gas had an important ecological purpose, the EPA has worked to all but erase it from the air. In this chapter, sulfur's importance to plants will be discussed. In Chapter 10, how sulfur keeps animals and humans healthy will be addressed.

Like many statistical numbers used in recording progress, the charts on the EPA website can be interpreted in multiple ways, A few of those representations are included at the back of this book. Regardless of how the information is reported, the EPA claims success for reducing sulfur dioxide levels by as much as 92% since 1970. In other words, the tested air has less than one tenth of the ambient sulfur it had when the Clean Air Act of 1970 went into effect. As a gauge of what is defined as air pollution, the number indicates a high level of success. In a real world application, however, that number may not tell the whole story.

In considering its impact on plant and animals life, readers need to remember that the current population for the United States is 50% more than it was in 1970. These reports also fail to consider the sulfur needs for an increasing food supply. For rural locations, EPA testing would seem to indicate a tiny fraction of sulfur available for a growing number of organisms which need it.

As sulfur dioxide and other gases have been reduced, the discussion of climate change has become louder and louder. Is this coincidence or are the two connected? The absence of sulfur in the atmosphere makes far more sense as a contributing factor in climate change than an over abundance of carbon-based gases which are readily produced by the human body. The next sections will provide evidence for such a conclusion.

Sulfur's Partnership with Nitrogen

Farmers have known about the connection between nitrogen and sulfur for some time. Without sulfur, plants fail to develop the vibrant green color that is an indication of healthy growth. Growing nutrient rich foods can be difficult without adequate sulfur. However, the role of gaseous sulfur may have been seen as a minor component rather than a major one.

Tobacco plant deficient in sulfur (USDA)

Called an immobile element that is also nitrogen fixating, sulfur serves the plant by remaining in the leaves rather than circulating throughout the plant as other nutrients do.

As a gas, sulfur dioxide is absorbed through the plant's pores and grabs nitrogen from the air. The nitrogen combines with carbon dioxide and water for growth. Sulfur remains in place until the plant is harvested or dies. If harvested and used for food, the sulfur is available to animals and people. If the plant dies, the sulfur returns to the soil where it remains until the next plant needs it.

This simplified version of the sulfur cycle illustrates the completely renewable nature of the environment. Nothing is ever wasted nor without purpose.

A sulfur imbalance happens when ambient sulfur levels drop too low during periods of plant growth or stress. Generally, there is enough ground sulfur to start the plant but not always enough in the air for the new growth to survive. This is where computer models blame radiation as the cause of wilting or dying plant life. Could the answer be sulfur deficiency rather than sunburn?

Produced daily on a global level before the advent of electricity, by-product filled smoke from organic and fossil fuel use released gases into the air to maintain low sulfur levels. Is there something else that needs to be considered here that makes this about more than just sulfur in the air?

As much as people want to think of the environment as something that needs tending, it is fairly autonomous. Without human intervention, wildlife has its own system of recycling sulfur through death, decay and rebirth. In this way, there is a balance between sulfur which is part of living tissue and available sulfur in the air and soil. The "just enough knowledge to be dangerous" tendencies of mankind has altered that natural cycle.

By living in close proximity, cheating death with medical treatments and creating a overabundance of possessions for his use, man has changed both the space and time frame over which sulfur and other elements are naturally recycled. Until people decided air had to be clean to be healthy, air pollution provided the one method by which the environment could still do its job. Without reverting to a gaseous state, elements such as sulfur, nitrogen and carbon fail to circulate to where they are needed. Without air pollution, plants and animals struggle to thrive.

Urban vs. Rural: The Danger of Equal Treatment

What is it about an educated human race that seems to believe equal treatment is always better than differentiation? The United States was founded on a principle of equality, while at the same time resisting any provision by which equality is mandated. If equality is an abstract goal but undesirable reality, why would government gauge urban and rural communities as environmentally equal. Obviously, they are different. One houses huge populations while the other feeds them. One is rendered helpless by any disruption of the power grid and one keeps going because they are prepared for such events. But with a stroke of a pen, federal, state and local governments put into place equal but arbitrary environmental restrictions that negatively impacted the financial and physical well-being of rural locations.

Between the end of WWII and the early 1970s, there is no doubt that care for the physical environment was a low priority. Weather, however, was moderate and predictable when compared to the period between 2000 and the present. Healthcare expenses were low and communities were economically stable because of a decentralized economy that produced air pollution where it could most effectively be used.

Life changed when the EPA targeted five common gases for regulation. Mandating expensive filters and collection methods, small rural factories could not compete in the cost analysis column. Operations closed but left behind the low paying service industry jobs and high unemployment rates that soon dragged down the standard of living and tax base.

In contrast, urban manufacturing got bigger, the wealth gap widened and weather became spastic. The solutions that the EPA recommended as a way of preventing poor health and smoked filled skies, instead, unintentionally caused an even greater concentration of unwanted air pollution in urban areas. In turn, rural and farming areas that could better handle such gases were left without good paying jobs. Farmers who put food on the table for those with high paying urban jobs dealt with a sulfur deficient air mass that meant high production costs but not higher market prices.

Human Migration to the Poles

As a thought exercise, consider how other countries are faring in the midst of drought and global climate change. If the United States is approaching high drought levels, what happens to smaller countries without mega farms and interstate highways to transport food.

One example is Ethiopia. In light of the information presented here, can the world stand behind its assessment that this county--and others like it--has experienced decades of drought randomly and without cause? Does it not seem more likely that when electricity modernized this region of the world, it brought with it the same weather disrupting pattern that existed in more moderate climates. Was Ethiopian air already so pristine that without fossil fuels, it could not produce so much as a gentle shower?

According to the World Health Organization (WHO) website, as much as 40% of the world is experiencing some level of drought. With an estimated 700 million people in those drought prone locations, what is an inconvenience for the US is far more life threatening to smaller countries.

With no solution in sight, these countries are prone to political instability.

The migration of the human race to the cooler regions of the world has begun. As a result, this exodus is leaving huge areas of land fallow and forgotten. Who will remain to put the local ecosystem back on track as civilized countries welcome immigrants to their borders? Which is more important: to offer political asylum to a few people or help stabilize a politically ravaged country by helping it solve its own climate change problem? Foreign policy that focuses only on political struggles will do little more than hamper global climate change recovery.

Population numbers or even population density is not the issue behind immigration in the United States. The matter is sustainability. The US ranks about 145th in how crowded the country is. The nation has the room for more people, but what is the ultimate cost to the global environment when poorer, struggling countries are environmentally abandoned because of political rhetoric?

One question that needs to be answered is *how does a global economy fit into the concept of environmental sustainability*? If products are not produced where they are needed, if regions cannot produce enough food to sustain themselves and if weather is impacted by environmental policies made in neighboring countries, how can any country live up to its pledge for environmental stability?

For much of the last fifty years, this country's governments (federal, state and local) have been obsessed with economic growth while science has attacked global climate change as it it was taking a magnifying glass to an elephant. Both ignored the big picture but took the lead and asked the world to follow their environmental example. Bolstered by its success on the battlefield as it championed democracy around the world, the US became overconfident in its ability to solve every global problem. In its desire to be a world leader, the United States instead, may have brought the world to the edge of a climate change cliff.

Sulfur and the Food Supply

How food is produced for today's market and how its nutritional value is reported could take up a book of its own. Suffice it to say, like scientific data, current information about food and nutrition is largely inaccurate. What happened to produce this maze of assumptions about food started even before the Clean Air Act of 1970.

In the late 1960s the federal government gave food producers more leeway in labeling their products. Before then, food testing was common but its use slowed manufacturing of new products such as Pop-Tarts, flavored chips and sugary cereals as well as multi-flavored taffy-style candies. Instead of real-time chemical analysis, producers were allowed to substitute mathematical estimations of food value based on the recipe.

On paper, it all seems quite good but in reality, the figures on which these calculations are often based come from a food database that was created in 1966. During a peak period of air pollution, foods tested in the 1960s would have likely been higher in many common minerals such as sulfur, magnesium, calcium and others. Blogs and other medical advice based on current guidelines for a healthy lifestyle may be little more than a situation of smoke and mirrors as companies play switcheroo with ingredients to make products more cost effective.

Vintage Food Label - Date Unknown

Today, there is no way to tell if nutrition data determined mathematically is accurate. Beside the obvious difference in numerical calculation and chemical analysis, guidelines give producers as much as a 20% margin of error so that labels can be printed before foods are even prepared. Labeling of prepared food from restaurants and self-serve delis are equally erroneous.

The take-away here is that no food label should be considered a fair and accurate representation of nutritional value. Similarly, charts and dietary recommendations found on the Internet may be largely based on data that could be as much as five decades old. While some farmers conscientiously test their soils and fertilize accordingly, not all do. Food value is dependent on the environment in which it is raised. Catch phrases such as "farm raised", "grass-fed" and "locally grown" are more to satisfy public impression than to guarantee food value.

Special considerations should be taken when preferring foods listed as "organic" over traditional foods. In order to gain organic designation, lands must be away from areas of high air or water pollution. If that equals low sulfur levels, organic products may have even less food value than traditional. Imported foods are also questionable since there is no global benchmark for nutritional value.

Review: A Timeline of Climate Change

Before heading into the impact of sulfur deficiency on human health, a brief review of the timeline for climate change is in order.

While science has identified multiple cycles that are known to keep Earth's ecosystem renewable and efficient, much of that knowledge became available after the global adoption of electricity. Using that information to look at history, it can now be inferred that the use of electricity disrupted environmental cycles that had been in place for centuries whenever the burning of fossil fuels and other organic matter was used by humans. Without gases such as nitrogen oxides and sulfur-based compounds along with dust and smoke particles, clouds could not form with enough stability to draw moisture over mountains and large expanses of land. The United States experienced a six-year period of drought that added to a global depression in the 1930s.

Focus on sulfur based drugs and nutrition in rural areas in those years might have been a matter of coincidence but they support a hypothesis that sulfur and other micro-nutrients were somehow missing in the environment during that time.

With increases in the popularity of the gasoline powered engine, weather returned to normal and people seemed healthier and happier. . Without cultural moderation, environmental pollution became commonplace leading to the conclusion that air pollution was an undesirable and unnecessary environmental feature.

With the adoption of clean air policies on a global basis, the world is again in a place of sulfur deficiency. Not only are states experiencing drought conditions as government energy restrictions fail to produce particulate matter for cloud formation but the US is experiencing a nationwide health crisis as diseases that were fairly rare fifty years ago have multiplied. Like the rise of sulfa-drugs in the 1920s and 1930s today's pharmaceuticals frequently include sulfur, nitrogen and oxygen in their formulas. Again, this may be coincidental but it does not take much of a leap to conclude that the medical industry is fully aware that air pollution is not the cause of these health ailments.

Rather than scientifically tracing these occurrences back to find a cause, the Internet's lopsided information system steered the public into believing in a 150 year old theory that had been proven incomplete more than a century ago. Now, leaders and advocates double down on carbon-based gases as the cause of climate change rather than investigating their own thought processes.

A return to moderate levels of well dispersed air pollution is the answer but until leaders are willing to roll back EPA regulations, restrict the use of gases that maintain the upper atmosphere and decentralize its economic base, climate change is out of the reach of possibility. In the next chapter, the extent of sulfur's impact of health will further explain how the US government compromised the health of its citizens as a trade-off for economic superiority.

10

Sulfur and Human Metabolism

As modern healthcare is considered here, the first question that comes to mind is ***How did the species survive if the human body has as many problems as healthcare claims?***

Even skipping over the millions of years about which scientists can only speculate, there are roughly 5,000 years of recorded history that document man's ability to survive in less than perfect conditions. If the human race had not been designed to thrive in his environment, and Darwin's theories on extinction are even close to accurate, humans should have been an extinct species long ago. Even with all that has become healthcare, the fear of human extinction looms as part of the climate change discussion. Is the human body and its environment that poorly matched that civilization must create a protective bubble in order for people to survive? Is it not more likely that mankind has created his own chaos and chosen to live in less than optimal conditions that he himself designed?

What could have possibly happened in the last century and a half that turned scientist's attention toward preventing illnesses? Certainly, the rich had the means to make discoveries without economic gain. There were also those who were willing to support the work of the scientifically inclined as a means of gaining status and there would always be those who craved immortality. But, why would there have been a focus on health if there had not been a drastic increase in illness and death? What would have been the purpose?

To be clear, in this discussion, there is a line drawn between the practice of medicine which minimizes suffering and the scientific goal of preventing death. Environmentally, a world without death is one that becomes stagnant and unproductive. The *Circle of Life* is not only something that is addressed in children's films and Broadway plays. Today, the obsession to keep things alive has taken on an irrational urgency that makes little sense when preventing death does not translate into living a healthy and rich life. Is there something that science has known all along that it is trying to correct? Are there other reasons for these discoveries that have nothing to do with personal well-being? As has been a major theme throughout this discussion, the role of economics and political ambition cannot be ignored when understanding how human health and healthcare have changed.

The Rise of the 'Maybe' Drugs

For decades, the pharmaceutical and healthcare industries have promoted preventative healthcare. What was once a doctor's advice to eat well, stay active, avoid extremes and enjoy life has since become a regimen of multiple medications that often come with undesirable side effects.

In 1960, healthcare amounted to five percent of the U.S. economy (brookings.edu). Today it is approaching a quarter of all national spending. In similar comparisons, the average expenditure for health care in 1961 was a mere $154 per person. As of 2018, the annual cost of care per person had risen to nearly $12,000 (thebalance.com).

In large part, the rise in healthcare costs came from what this book will call **_Maybe_** drugs. Named this because "maybe" a patient will need them and "maybe" they will extend life, these medications were designed to combat the possibility of illness, not illness that had actually occurred. Blood pressure medications were given to prevent stoke. Blood thinners were given to prevent blood clots. Statins were prescribed to rid the body of dangerous levels of cholesterol and ward off heart attack. Then, diets changed to low salt, high carbohydrate menus to keep weight at a reasonable level. Age, failures of the food industry, genetics, evolution and even living longer were reasons that were given to patients as doctors prescribed more and more drugs. Could all of these conditions be linked to one environmental misstep taken by an overconfident and barely knowledgeable group of government leaders and advisors? The evidence says, "Yes!"

Ironically, the use of _maybe_ drugs began in the years between the adoption of electricity as a power source and the coming of age of the automobile in the late 1930s. Many of these medications were based on sulfur. Penicillin (antibiotic), insulin (diabetes), and metamizole (pain relief) were all discovered following the Spanish Flu pandemic (1918). Losing nearly one third of the world's population was reason enough to look to new ways of treating illness.

Is it coincidence or correlation that had sulfur-based drugs discovered during a time of reduced fossil fuel use? Is it coincidence or correlation that those drugs are no longer recommended because they interfere with new formulas which also include sulfur as an ingredient? As the US approaches zero levels of ambient sulfur, is it also coincidence that many of the new drug formulas coming on the market include sulfur and nitrogen in their chemical makeups?

Today, there is a quietly growing industry of homeopathic and over-the-counter supplements that contain sulfur and nitric oxide. While health claims are not implied here, the evidence that human metabolism needs a regular influx of sulfur and nitrogen seems evidence based. If the health industry knows about the importance of sulfur and nitrogen in human health and agriculture understands the importance of these elements in producing a nutrient rich food supply, then why is the federal government restricting their release into the atmosphere? What is government protecting--its image, the urban lifestyle or a healthcare industry that might not have even been needed if government had not intervened in environmental concerns?

Sulfur and Naturally Preventing Human Illness

To be clear, the information presented below is done on the basis of chemical formula, not medical advice. Sulfur is a powerful element. While it is estimated that a 200 pound person retains about a third of a pound of sulfur, how quickly sulfur is metabolized and how much is needed on a daily basis varies from person to person. Unlike medications, vitamins and minerals ingested through food provide small, continuously available levels of important nutrients rather than large doses that are difficult for the body to regulate. By restricting sulfur and other fossil fuel-based gases, the Clean Air Act may have unintentionally subjected millions of Americans to a lifelong sulfur deficiency that caused years of suffering and drained more than a few bank accounts.

The link between sulfur deficiencies and specific diseases is becoming more apparent. The list that follows explains those conditions that make up the most common and best documented evidence of such a link. Where possible, chemical formulas follow the medical term. "S" denotes the presence of Sulfur and the number immediately following it represents the number of atoms in each molecule. Letters without numbers denote one atom of that element in the formula.

Obesity
Lipoic Acid ($C_8H_{14}O_2S_2$) is currently sold as a dietary supplement for weight loss and obesity control. This compound occurs naturally in

the body given a sulfur rich diet. Several recent studies link it to reduced weight and increased metabolism. Since it occurs naturally, FDA approval is not needed for such products.

Diabetes

Insulin ($C_{257}H_{383}N_{65}O_{77}S_6$) as a drug was first formulated in 1922. Naturally, it is formed in the pancreas and used to break down carbohydrates. Without sulfur the body cannot make its own insulin. What is now known as pre-diabetes, insulin resistance or insulin insufficiency may be related more to a low sulfur air mass than a dysfunctional human body.

Celiac Disease

Little is known about the sudden rise in celiac disease and its causes. Of interest is the growing number of calls for US farmers to fertilize grains with sulfur. In this way, grains would have a built in supply of sulfur that would spur slow metabolism of complex carbohydrates and encourage a steady production of insulin. Wheat and other grains grown in sulfur deficient conditions would be more difficult to digest. With the USDA and environmental activists recommending a more plant based diet, it is not known if adequate levels of sulfur can be obtained through food alone. Ambient sulfur dioxide may prove to be the reason people living in poor countries are able to live on a diet consisting mostly of grains.

Blood Clots(Lung and Deep Vein)

The body's natural anticoagulant, Heparin ($C_{12}H_{19}NO_{20}S_3$) can be produced by the body in sufficient quantities when a sulfur rich diet is available. Heparin keeps blood circulating even in

Grains ready for harvest

times of dehydration and injury. Heparin should not be confused with Vitamin K which helps blood clot. Vitamin K does not have sulfur in its chemical makeup. The most common medication to treat blood clots (Warfarin) does not contain sulfur but a new medication (Xarelto) does contain sulfur.

Thiamine - Vitamin B1

With a chemical formula of $C_{12}H_{17}N_4OS+$, Thiamine aids in many metabolic processes but may be specifically related to the nervous system and brain health. A Thiamine deficiency is often found in alcoholics likely due to the high level of carbohydrates they consume. Thiamine may be used in the production of insulin when foods devoid of sulfur are eaten such as those containing processed sugars. It should be noted that the addition of Thiamine and other B vitamins has been supported by the USDA since the late 1930s. Unfortunately, as the public grapples with unexplained health issues, people abandoned enriched grains for products marketed as whole grains--not always with better results.

Biotin - Vitamin B7

In the same way as Thiamine provides a specific purpose, Biotin ($C_{10}H_{16}N_2O_3S$) supports the breakdown of food as a source of energy. While it may be a backup system for the body's insulin supply, Biotin's greater purpose is to support the production of hair, nails and skin tissue. As the body's largest organ and greatest protection against disease, the skin depends heavily on a good supply of sulfur for maintenance. A new medication (Dupixent) for acute skin conditions not only has sulfur in its formula but also nitrogen and oxygen.

Human Sulfotransferases - Multiple Formulas

Functioning in very specific ways, sulfotransferases are enzymes that take on important roles within the human body. Their purpose can be as varied as protecting the body from infection to speeding up metabolism and moving nutrients from one part of the body to another. As enzymes, sulfur may be substantially more important to overall health than first thought. Its use in so many enzymes makes it difficult to determine how much sulfur is needed on a daily basis.

Chondroitin Sulfate - Cartilage Maintenance

As thousands of individuals each year have knee, hip or shoulder replacement surgery, the question has been "What happened to the cartilage?" Like many other biological processes listed here, cartilage repair requires sulfur and without sulfur there is no cartilage. The assumption that as people age, the body simply stops producing cartilage may not be accurate but be related to a dietary lack of sulfur. Not only found in major joints, cartilage degradation can affect the voice and larynx as well as bronchial tubes and breathing. Well known facial features such as the nose and ears are comprised of cartilage while the same material keeps spinal discs moving smoothly. Over-the-counter supplements with MSM (Methylsulfonylmethane) are common for treatment of this condition. Of interest is a condition known as *Psoriatic Arthritis* which involves patchy skin dryness and swelling of painful joints. One of the latest drugs, Otezla, contains nitrogen, oxygen and sulfur.

Eye and Lens Health

Like the ears and nose, the lens of the human eye is constructed of a cartilage-like substance. As far back as 1967, researchers express concern that not enough was known about sulfur and its function within the eye. A lack of sulfur may contribute to cataract development and retina damage. A specific protein component, methionine which is the most common sulfur-based amino acid tends to be found in the muscles that control the eye's ability to focus automatically.

Without capillary action, the lens must get its nutrients from other means. The old song, **Smoke Gets in Your Eyes** may have new meaning. If the lens can absorb nutrients from the air, air pollution provides the most direct source of sulfur dioxide to the lens. According to the National Eye Institute website, the incidence of cataracts is growing rapidly with a 20.5% increase between 2000 and 2010 alone. Other eye conditions, attributed mostly to age rather than environmental conditions, have also increased during the same time period.

Human Reproduction

A brief review of studies and informational websites brings with it a wide variety of conclusions. In general, studies see air pollution as a negative but often see sulfur as a substantially beneficial mineral. Likewise, studies of methionine, a sulfur containing amino acid, conclude it as necessary for good fetal development but also a cause of infertility. This study area is one that has largely been done after sulfur dioxide levels had reached near zero levels. Is it air pollution that is the cause of infertility or is it a lack of sulfur dioxide that is a contributing factor for the steady increase in such conditions?

Often, printed reports of these findings do not go far enough to explain how and why such results occur. Even with the ambiguous nature of its research studies, the impact of sulfur on human reproduction is clear.

One detailed and collaborative paper originally released in 2003 (*Sulfur containing amino acids and human diseases*, Pub Med) looks across several specialties and explains the confusing results. Methionine is broken down to make other sulfur-based amino acids and enzymes as well as insulin. According to this report, it is when the body substitutes a non-essential amino acid for methionine that difficulties with human reproduction arise This information can be interpreted to mean that when the body substitutes a less complex molecule in muscles, organs or skin, it signals that the environment is not capable of supporting additional life. Rather than air pollution being the cause of infertility, it seems as likely that it is the lack of air pollution that impacts reproductive rates.

Additional studies and medical advice sites speak of the clear link between enzymes that contain sulfur and the balanced production of reproductive hormones. Common drugs for males such as Viagra and Levitra contain sulfur. The number of males worldwide experiencing performance issues has risen steadily from the estimated 152 million patients in

1995 to almost double that today (hopkinsmedicine.org). Again, the link between a lack of atmospheric sulfur appears to trigger low conception rates in both men and women.

Alzheimer's and Dementia

Over the last decade, the presence of incomplete protein-like masses in brain tissue of Alzheimer patients has given rise to a different type of preventative drug. Donanemabe by Eli Lilly is a large molecule compound with nitrogen, oxygen and sulfur in its composition. Its competitor with a nearly identical formula, Aduhelm (aducanumab) by Biogen recently achieved accelerated approval from the Food and Drug Administration sparking a debate of the FDA approval process. Could these partial protein molecules that seem to collect in the brain be the result of breathing air that no longer contains nitrogen and sulfur based gases? How likely is it that these patients would see a similar level of improvement by supplementing their diet with sulfur and nitric oxide?

Historically, Alzheimer's was first identified in 1906 by Alto Alzheimer, with his study being limited to his examination of hospitalization records and brain tissue of only *one* patient (verywellhealth.com)

Like Greenhouse Gas theory, Alzheimer's work did not make it into medical culture until 1995 after former president Ronald Reagan was diagnosis with the symptoms. Could this condition that has devastated countless families have been preventable if a nutritional basis for it had been considered? Instead of grasping onto an out-of-context theory, medical adoption of Alzheimer's very limited research spawned millions of dollars in pharmaceutical research and bankrupting residential care.

As for the new drugs, with a reported 32% effective rate over the short term and an estimated cost of as much as $70,000 annually, sulfur and nitrogen may be the reason these new drugs work at all. Currently available sulfur and nitric oxide supplements can be purchased of less than $50 per month. Because they are known to be needed by the body, the FDA does not reguire study before placing them on the market.

Methionine: An Essential Building Block for Life

In the study of nutrition, explaining what a protein is can get confusing. Often used along with the term "amino acid," proteins are made from these smaller molecules, not equivalent to them. Both amino acids and proteins are divided into two categories. Amino acids are considered essential or non-essential. Proteins are labeled complete or incomplete. Essential amino acids must be ingested as food with non-essential amino acids manufactured by the body for specific uses. Similarly, complete proteins must come from food, generally believed to be animal sources, and incomplete proteins come from a variety of plant and animal sources.

Methionine, cysteine, homocysteine, and taurine are amino acids that contain sulfur. Only methionine is considered an essential one. That means, methionine becomes one of the body's best sources of sulfur once ingested.

In order for the body to utilize sulfur, it must be part of a molecule that the body recognizes. Methionine is a molecule the body recognizes and can put to use without recreating it. Known to provide flexibility to blood vessels, heart muscle and skin as well as repair liver damage from drug and alcohol use, methionine levels can be hard to maintain in a sulfur deficient atmosphere. The concern comes when the body must break down methionine in order to produce insulin or other important hormones that use sulfur.

As nutrition has become more of a medical factor, all manner of judgements have been made about how sulfur-based amino acids perform. Some studies discount the importance of essential amino acids in support of a plant based diet. Other reports like *Sulfur-containing amino acids and human disease* mentioned previously, show strong connections between a number of health conditions and a shortage of sulfur based amino acids. In addressing chronic illness after trauma such as long-term

COVID-19 symptoms, the report clearly links deficiency of magnesium, calcium and sulfur as a primary factor in long recovery times.

The Take Away for Rural and Disadvantaged Localities

A century ago, the American Home Economics Association), now known as the American Association of Family and Consumer Sciences (AAFCS) partnered with the USDA and its Cooperative Extension Agencies to mount an extensive campaign to improve nutrition and health across the country. Focusing on rural and poorer neighborhoods, this work was instrumental in raising the standard of living for many families. By showing them how to grow, preserve and prepare good quality food, these efforts helped lower the incidence of childhood malnutrition and death. Nutrient deficiencies such as Goiter (iodine) and Rickets (Vitamin D) were largely wiped out with simple legislation that added these nutrients to commonly used products like salt and milk. The process of fortifying grain products with iron and B vitamins (some of which contain sulfur) also occurred during this time period, a sign that sulfur deficiency may not have yet been identified but that its effects were being felt.

Much of what is now known about nutrition came from this comprehensive and well respected effort. Now, the "natural food" movement has the public questioning the validity of these practices. Believing they are doing right by the environment and their health, the public as a whole runs the risk of developing bone and sight weaknesses when they use plant-based milks which are not required to include Vitamin D & A. Similarly, gourmet Sea Salt is used instead of Iodized salt which keeps the thyroid healthy and is linked to breast cancer prevention. The name, Sea Salt alone, does not indicate iodine in its formula although that is a common myth. As sulfur has disappeared from the atmosphere, people, especially those who live in pristine country air, have to work harder to stay healthy. In the Informational Chart section at the back of this book, two color coded US maps indicate the large swaths of land that are considered free of all sulfur dioxide. Is it any surprise that government had to throw millions of dollars into health care for these areas--all to make up for what the environment and fossil fuels would have provided naturally.

To add more fuel to the nutrition debate, those who write wellness advice columns tell people that taking vitamins and using fortified products are a waste of money. While likely coming from authors who live in areas of moderate levels of ambient sulfur, a single standard for diets and medical care can no longer be accepted.

Home Economics Class - 1931

If nutrition was a problem for isolated communities as far back as the 1930s, why was sulfur not identified as a need then? The answer is simple. High sulfur coal was still produced and used throughout many areas up until the EPA restricted sulfur in the early 1970s. Not only did the country lose a good environmentally beneficial heating source, fine homes with central heating that were fueled by coal, were often torn down because of the cost of retro fitting a new heat source. Not only did these communities experience an environmental hit but a financial one when regulations made it difficult to keep and maintain homes that had been in families for decades.

Here is where rural communities and depressed urban neighborhoods find common ground. Left to deal with the outcome of decisions made to benefit locations where economic growth is the focus, rural and poor urban neighborhoods have been short changed economically as well as losing air pollution which aids plant growth and helps maintain human health. A second look at the chart mentioned above, shows small light white and yellow patches where population density is higher and sulfur levels have beneficial effect on health. Rather than provide economic opportunities that are also environmentally friendly, the government provides unlimited healthcare to treat conditions government action may have put into motion. Without considering the impact on disadvantaged locations, the government unintentionally set in motion the very health crisis it had claimed to avoid when combating air pollution. The only difference is that instead of treating patients for diseases from too much air pollution, providers now treat them for nutrient deficiencies related to too little air pollution. Why is moderation so difficult for government to manage?

11

Tweaking Expectations

Before this discussion can be closed, there is a sensitive subject that has to be addressed. Without recognizing human nature's role in this situation, the information here could be the beginning of a dozen or more conspiracy theories as to how government railroaded an unsuspecting public for its own gain. Certainly, there have been elements of greed that have come into play but no amount of greed could have sent the world spiraling into an emotional tailspin faster than its own fear. The greatest danger of a faithful and educated world is fear. Fear of what it knows. Fear of what it does not know and faith in something more than personal responsibility and consciousness. Until individuals understand that life is a participatory experience, no amount of faith or knowledge can return the environment to its original condition. The world's citizens will also need to reconcile their image of life with one that is environmentally sustainable. People are the stewards of the Earth and it takes their attention and effort to keep it from being compromised.

How does one change what people believe when it is all they know? Critics say it cannot be done and will make dire predictions about the future. Visionaries believe anything that puts the responsibility for disaster at someone else's doorstep and expects a quick end to a nasty situation. But there are the realists in the world who have already drawn similar conclusions to those outlined in this book. They could bring about a methodical and simple solution to environmental crisis except for one thing. Those who fear risking the status quo will resist action and those in positions of authority will keep them fearful so as to maintain power over them. Will strong leadership emerge to change fear into confidence?

The present day environment has been in place longer than those currently living on it. To them, it is considered normal even though it makes up only 1/40th of recorded history. The post-electricity environment is not a normal one. Electricity is not going anywhere but it expanding it should not be what is recommended. Like the belief that the world is flat, science decided air pollution was the danger. Failing to look pass the conditions it could see, science has resisted looking for the reality of a climate change cause that was always just beyond the horizon. Will fear of admitting to its unintended miscalculations keep science and government locked into decisions that can really compromise the planet or is it time to put fear aside and do what scientists are supposed to do? Find the truth.

The responsibility for finding a solution to climate change does not end there. The world's population will need to step up and participate in the decision-making process. Decisions, after all, determine the future, not computer models. Government is not an inflexible and dictatorial body unless citizens shirk their responsibility to hold it accountable. Activism does not require protests, attacks on government buildings or the spread of misinformation. All it takes is rational thought and personal commitment.

Listed below are just some of the ways that individuals can become participants in environmental recovery. With a word of caution to stay within current legal limitations, individuals can have a greater impact on climate change by enjoying life than by fearing the results of inaction.

Choosing a Local Model over a Global One

There is nothing that makes a community feel stronger or be more confident than local success and there is no more powerful tool then for family households to spend their dollars locally. No amount of government support, special events or tax breaks can match the steady influx of cash and customer loyalty needed to build a sustainable and local economy.

Equivalent tos casting a ballot for environmental policy, choosing local is not running to the big box store and grabbing up the imported bargains, nor is it buying online (unless its curbside pickup). It is choosing to buy less, pay more and being flexible enough to support a small, but local, manufacturer.

Choosing local can also mean expecting more from utilities and suppliers. Many times a good argument can be made for installing smaller backup facilities that operate during major outages. Backwards power generation using fossil fuels from the end of a line seems counter intuitive but its cost may be well worth it when calculated against lost revenue and increased repairs. Jobs should not be the only criteria for businesses to meet. Servicing local markets first is the best incentive to get reluctant customers to the door.

The world has tried a global economy, and while it brought cheaper products, it also brought environmental issues. Buying local might not be as trendy but it is sustainable.

Rethink Healthcare

As the Internet has been forced to alter its search engine parameters and justify its focus on advertising revenue, a steady stream of evidence is becoming available that documents the importance of sulfur in human health. Many of these articles and reports have been catalogued on the

National Institute of Health's website. While the agency absolves itself of any obligation to use or support such information, there seems little question that at least some well-placed medical professionals know that clean air policies have put individual health at risk. Regrettably, the nation's focus on good health has turned the industry into a business rather than a calling. What could the US do with all the money its now spends on healthcare if a simple solution to it is found?

Equipment for self-monitoring of vital signs has become readily available and produces more relevant results than those produced in office visits. A high potency multi-vitamin that includes recommended or better levels of B-Vitamins should be considered as needed until it is known how climate change will end. Most of all, expect more than a one-and-done, ten minute questionnaire based evaluation of health status. Like the weather, health is ever changing and should be looked at as a pattern not a single event.

Rethink Recycling - Burn, Baby, Burn

Recycling of paper was virtually unheard of before the EPA established its recommendations for best practices. Older buildings were fitted with incinerator chutes to burn trash rather than save it. It was common for rural locations to have a 55-gallon drum barrel at the back of the property where trash was burned late in the evening after the breeze had calmed. Plastics were not common and were saved rather than thrown away. All of this is to point out that three decades of recycling efforts have not impacted the climate change trajectory. If it hasn't worked in the last thirty years, it is not likely to make a difference in the next fifty.

From a scientific perspective steady incineration of paper products and natural fibers would have the potential to slowly increase levels of ambient sulfur and put cloud forming particles into the atmosphere. It seems worth the experiment to see if methodical incineration of the tons of paper products used by California's largest cities could seed the atmosphere and produce consistent and steady rain over the Sierra Nevadas. Much better than allowing nature to handle the job through massive wildfires, would a year long step backwards be any less disruptive than legislating massive construction projects to divert even more water from natural sources?

Expect Government to be Different

As individuals use their financial voice instead of relying on public complaints, elected officials lose their footing in controlling the narrative. Rather than accepting the letter of the law and settling for a quick public comment session that is often ignored, those in government should expect to educate their citizens and be responsive to their citizens educating them in return. Making decisions in order to appease the few is how the climate became compromised. That lesson has been learned. Instead of an economic arrangement, this is an honest give and take between those who make the decisions and those who must live with them.

Replacing public-private partnerships, programs can be put into place to use industrial parks and downtown spaces to support micro-businesses without preconceived mindsets or limitation on what is and is not a viable business. An owner that wants to open a small museum or gallery is given the same level of support as the fresh food cannery which will hire fifty people. Changes made to the old department store that had closed thirty years ago could potentially support a hydroponic fish farm in the back of the building and a quaint cafe in its storefront. With 2500 sq.ft of container grown strawberries on the roof (roughly producing a ton of fresh strawberries each year) a building's environmental footprint becomes far more than how much water it uses and how much trash it creates.

Expect Foreign Policy to be Help not Leverage

With economies shrinking to focus on local sustainability, food exports and imports are no longer a tool for leverage. Even in third world countries, new technology spearheaded by one-person food operations and locally grown initiatives allows communities to produce marketable products. As air pollution returns so does the rain that makes it more likely for countries to feed themselves. A secure global economy is not one based on trade and industry but on feeding the people within each country's borders. What was once a goal of the Peace Corps and charitable organizations, teaching communities to survive within environmental limitations replaces administratively bloated food programs that create dependence instead of continuity. Without abandoning the unique culture and values of the region, products are those that people want, not those that are surplus a world away.

As people feel secure in their environment, the need for war dwindles. Revolutions are not started by people who had enough to eat, a roof over their heads and a source of adequate income. Take away the need to fight and you take away the fighting.

Cites Consider Neighborhood Style Ecological Renewal

Urban locations will need to address population density before air pollution is manageable. Adopting a neighborhood model with appropriate green space may only require a reassessment of housing needs. Are all the buildings in a district at 80% capacity or more. Can near empty office buildings or hotels be adapted to mixed-use structures so people can live and work in the same building. When buildings are demolished can the owner choose to put in a park and "sell" his development rights in the way rural landowners are compensated for leaving their lands in a natural state. Cities have grown up with a layout that came from Europe 500 years ago. The boroughs of New York City started as distinct neighborhoods, not wall to wall people. Why would returning to that model be anything but positive?

Rethink Agriculture

Agriculture, something of an air pollution sponge rather than a contributor, has been blamed for more than its share of climate change problems. Forced to specialize as local markets and processors move out of small towns, there is no reason that a new generation of farmers cannot turn two-acre lots and abandoned buildings into a supplemental revenue stream. With technology, farming does not need to be the 24/7/365 operation it seems. Hydroponic farming can grow healthy catfish and tilapia in a matter of months without the investments of building ponds and maintaining fishing boats. Poultry can reach maturity in a matter of weeks and can be processed for home or market with a minimum of equipment and space. All manner of vegetables can be grown and harvested for the fresh or canned market. Small micro canneries and processing facilities could fit into urban renewal plans with a bit of tweaking to the definition of a business. Mega farms could continue to grow for the urban and export markets, but expanding locally grown and processed options can have an positive impact of poverty and food needs.

Returning to the model that was used prior to WWII, Family and Consumer Science professionals working with young families can shift the image of home raised food from one of hard work to one of practical application. As people learn what their food has to do with their health, food labeling changes and a return to chemical testing makes labels more relevant.

Think Innovation Rather Than Accommodation

A return to a functioning ecosystem does not have to be the bitter pill it has been described. The world provides everything that is needed. The only accommodation to a healthy life is to stay within the limits of the environment.

What science calls pollution has been around for thousands of years.

Areas around shrines and temples in Asia are classified as areas of poor air quality because of incense use, not industrial operations. Yet, they experience less obesity and health issues than Americans. Daily burning of candles has been a part of other religions as well. If further evidence is needed to support sulfur gases in homes, it should be noted that a patent for a tin-can sulfur based scented candle was granted in 2006 as a way to rid the air of pet odors and other strong smells. What would the consumer choose-- a scented candle or an expensive drug regimen to control high blood sugar and weight?

Using sulfur compounds for cleaning is common in industrial settings. Sulfur dioxide has been used as a food preservative for dried fruits and beverages for years. How is it that sulfur dioxide in the air is a danger but sulfur dioxide in commercial food production cuts the chance of bacterial growth? Could the pandemics of 1918 and 2020-21 be a result of a lack of infection resistant sulfur gases filling human lungs? Studies with wood smoke are not uncommon but likely fail because they choose a treatment model rather than a continuous exposure model.

Rather than thinking about maintaining jobs for services that support unsustainable lifestyles, consider a higher minimum wage that is coupled with reducing the workforce. Would a return to the single income family be that much of an inconvenience?

Time to Pay the Piper

Inventors and politicians got the US into this situation. Why not expect them to get the country out of it? The Department of the Interior should become the new home of the EPA and, after a six month period of review, determine if the agency has a place in government or whether it should be dismantled. California's role in determining national environmental policy should be minimized as a result of long history of environmental abuses on it citizens and neighbors. Likewise, members of Congress from Northeastern states who presented the Green New Deal should be required to hold in-person town hall meetings in the South and Midwest so these clean air champions can understand the complexities and limitations of the legislation they propose. Solar and wind power should become elective sources of energy not mandated ones. Placed in the position of being accountable to more than their own constituents, it is the hope that politicians will realize they played a major role in bringing climate change to the world. Perhaps, in the future, they will support causes because they know what they are doing rather than supporting who will owe them a favor.

And then there are the inventors. Since the Clean Air Act of 1970 was signed, inventors have played havoc with the environment. Besides the damage done by the computer and the Internet to the information collective, incorporating semiconductor and cryogenic technology into everything electronic was impractical and environmentally harmful. The new rule for inventors needs to be "Just because it can be done, doesn't mean it should be." Patents, like drugs, should be awarded on the practical application of the technology rather than the ability of the designer to profit from it. If a teenager can understand that hot bacon grease will dissolve a styrofoam cup in seconds, it should not be that difficult to get a plastic milk jug to degrade in less than a thousand years. Inventions should make life easier not create world disaster.

One Problem Remains

The one kink in what can be a smooth recovery from over-the-top regulation of air pollution is the mining of gases for profit. Even if weather stabilizes and human health improves, the lack of cloud cover in the higher altitudes will make it difficult to stem the melting of the polar ice caps. Noctilucent clouds are their protection and lighter-than-air gases factor into their formation. As long as there is a reason for these gases to be bottled and sold, such trade will continue somewhere in the world.

Final Comments

What started ten years ago as research into how nutrition played into an inexplicable decline in human health is at an end. Not only providing the author with answers. this collection of information has unearthed clues as to how climate change began and how to solve it.

The clues were scientific bread crumbs left by frustrated researchers as well as over-confident "experts". Each path lead back to similar events and similar causes with the conclusion that something was wrong with the atmosphere. A fascination with historical architecture and a deep respect for the work of Madame Marie Curie raise quesions when greenhouse gas theory was considered. A pinch of farming experience, unanswered questions by doctors and a new drug that paired seeming unrelated conditions were what it finally took for the pieces to fall into place.

Beginning with the adoption of electricity, human preference for clean air has put the world in a situation of being deficient in sulfur along with its partner nitrogen. This nutritional void impacts every living thing on earth and under the sea. Even the variable nature of health and weather in different places and different times can be explained. It seems that the environment needs a balanced diet just like humans. It is only now that clean air initiatives can be seen as the starvation diet on which climate change occurred.

Some may think correcting this problem is just as simple as reversing the steps that got us here. Unfortunately, a quick return to massive air pollution is not the answer either. With individuals driving the recovery, and not government or business, a slow and sustainable level of air pollution is attainable. The world needs a healthy environment and *Air Pollution's the Answer!*

12

History Repeats Itself

The United States has been here before. Perhaps not in dealing with a environmental issue on the global stage but it has come to more than one crossroad that required a choice between defacing the environment and promoting economic growth. While individuals can make a difference in regional ecosystems, how this nation's government reacts this time is the one unknown left in meeting the climate change challenge.

Past history leaves modern researchers with two centuries of examples when government action pitted one group of people against another--all for the purpose of laying claim to land and resources that it wanted to control. The selective memory of the social justice movement ignores the reality that every race, ethnic background and religion were, at one time or another, robbed of their livelihood and heritage. Success in handling climate change depends on the American people realizing that race was only a tool to drive people to do what the government could not. To give reparations to all that lost something would bankrupt the government with a lifetime of debt and still not fix the environment. Climate change is just another situation of division being used to gain control. One group of voters want clear skies and another wants rain. Both could have their way if they would stop taking sides long enough to see that what both want is quite possible.

The Environmental War Between the States

The Civil War was not fought because of an overwhelming concern for a race of people who would be no better off after they were freed. It was an environmental conflict fought because the South was everything the North needed but could not have.

Within 25 years of its independence, the capital of this young country had been moved south, not once by twice. By 1800, the capital was located in Washington DC and much closer to Southern influences than Northern interests liked. Had the Civil War not happened, it is quite likely that the South would have become the center of commerce and trade for the country, with or without slavery.

The South had realized from the beginning that their lifestyle depended on a sustainable relationship with the land. Those who settled in the South achieved what the other regions of the country had squandered. The South had a decentralized and self-sustaining economy. Almost like small villages, farms and plantations relied on their own resources and the talents of those living on or near them. Towns were small and many. Churches popped up everywhere. Those who represented the outlying regions in state legislatures had as much decision making power as those in cities, perhaps even more. People dispersed themselves over large areas rather than congregating in crowded spaces which were known to harbor sickness and disease.

Without large scale industrialization, cities across the South were dwarfs compared to those in the North and Pacific Coast. As California joined the union, its southern regions decided to follow the example of the industrialized Northeast and go for growth rather than continuity. Of all US cities reporting in the 1860 Census, New Orleans was the only Southern city that ranked in the top ten. Its population, as a major port serving the Mississippi and Ohio River basins, was a quarter of what New York City was at the time.

As the country continued to carve up political alliances between North and South, Southern states became more and more independent. Regions produced everything they needed. Only surpluses were sold or traded. Only that which could not be made or found locally was imported. Already shifting away from individual labor, inventions such as the steel plow, drill seeder, threshing machine and cotton gin, minimized the number of hours slaves spent in the fields. Like today's migrant workers, cultivating and harvesting delicate crops were the only time extensive labor was needed. The image of slaves bent over short cotton plants every day of the year is misinformation put out to sway Northern sympathizers. With a sense of duty, land owners understood their responsibility to those who lived and worked for them. In contrast, workers in northern cities worked long hours in dark and dangerous factories without assurances they would make enough money at the end of the day to put a roof over their heads and food in their stomachs.

The South had everything needed to be a thriving nation of its own. Its many ports, which remained open year round, allowed trade with European and South American interests. Immigrants came to Southern ports but the South's balanced economy did not always have a place for unskilled labor The larger land owners understood the careful give-and-take between crops and residents. Sharecropping was possible but the land was always the priority.

Unrestrained immigration through New York and Philadelphia had caused overcrowding in most of the Northeast. With a shorter and cooler growing season than the South, Northern cuisine was based on meat, fish, dairy, dried beans and root crops with the occasional dried fruit or glass of wine. Growing fresh vegetables was difficult, even when the use of hotbeds extended the growing season. Compared to Southern cooking with its French and Cajun influences, Northern foods were plain but filling.

Homesteading to the West was as much about reducing population density as it was a need to provide food for the country's ever growing population. The North did not need the "freed" labor from Southern farms except as competition for European immigrants. The North wanted food, cotton for its textile mills, tobacco for its pipes and cigars and timber to build more and more housing.

The South held all the cards. There was only one way to grab the resources that the South had. Destroying its highly efficient culture would open up the system to those who sought to control it. The only way to undermine this group of states so that it played well with good people and politicians was to make it a moral argument against slavery. After four years of war, the rich agricultural legacy the North had desired was all but ruined. Families that had tended the land for generation lost it through quick claim deeds and tax sales. Former slaves had no recourse but to return to the fields except this time there was no assurances that the new land owner would provide for their needs.

The Quest for Land Continued

Like many other groups before and after them, slaves were merely a tool in the fight for economic superiority. Before the Civil War, it was the removal of the Native American tribes east of the Mississippi. Afterward, it was funneling those same tribes into what was considered unproductive lands. That lasted only until gold, silver and other valuable minerals were found on the reservations.

Between Native American relocation and the Civil War, businessmen would bring thousands of Chinese immigrants into the country to break the will of white European workers who had made mining successful and the railroad a reality. So great was the number of Asian immigrants during the 1800s that every major city of the day continues to have a neighborhood named China Town.

Even after securing all the land between the Atlantic and Pacific Oceans, the US was still grabbing for land. Alaska would be purchased in 1867 and rapidly settled when gold was discovered in 1898. A bigger and more environmentally damaging strike would come when oil was discovered in 1968.

How the Kingdom of Hawaii's fared under American rule is similar to that of Native American tribes. Settled by British and American profiteers in the early 1800s, the native population would be devastated by disease. Using a Southern plantation model of farming, Hawaii was turned into an export business supplying sugar to many other countries. To support this labor intensive process, large numbers of immigrants were brought to the Islands. After a fewl decades of cooperation with the Hawaiian royal family, outsiders would encourage constitutional changes that curtailed the powers of the monarchy. In 1898, the US government would broker a treaty with the new governing body of Hawaii. Going against royal advice, the island nation became a US territory in 1900. The new constitution and territorial status greatly limited the rights of the Native and Asian populations which had made the islands a lucrative source of income. Like treaties with mainland tribes, many of the promises laid out were never kept.

A Hidden Legacy of Land Surrender

One part of the environmental story that is rarely addressed is the formation of the United States' network of national park lands. Less than a decade after the Emancipation Proclamation was signed and for years extending through the 1930s, presidential authority, supported by Congressional and state action, commandeered large tracks of land "for the enjoyment of people." Covering more than 85 million acres, most Americans fail to realize that these lands were owned and maintained by citizens who had little recourse but to surrender their property and watch as it became a place of public recreation. Because these lands were obtained "legally" through government action, the true nature of how many US citizens were displaced is difficult to track. Even today, property can be legally seized for pennies on the dollar to support environmentally questionable policy such as pipelines and public utility right-of-ways for alternative energy sites.

What Happened to Government Of the People?

By the time Abraham Lincoln delivered his Gettysburg address and expressed his hopes that a "government of the people, by the people and for the people shall not perish from the Earth," he was likely aware that statesmen were anything but everyday people. Largely a body of retired military heroes, academics and career lawyers who had little knowledge of how to maintain anything but lobbying influence, is it surprising that the United States would grow into a country that attacked every problem as if it were a battle to be won? Unlike older more mature nations, the United States has yet to learn how to build consensus and adopt a national agenda that meets the intent of the Constitution not the letter of the law. Without continuity, the risk is great for the United States to repeat its environmental mistakes over and over and over.

The Battle Against Air Pollution

Like a campaign in war, the EPA treated air pollution as an enemy. There was only the goal of achieving blue skies and fresh air that mattered. To believe that climate change was anything more than collateral damage in this effort to create a perfect world is not supported by historical events. Before pointing fingers and setting up talking points, voters, as well as representatives on both sides of the aisle, should consider their role in this situation. Had it not been for a system that was focused on personal comforts and an ever higher standard of living, climate change would have been solved years ago. Climate change continues because it is more profitable to find theories which support the emotional rhetoric than to find the solution and solve the problem.

Unlike greenhouse gas theory, what has been presented here explains every major environmental issue without the need for computer analysis.

Based solely on science in a real world application and taking into consideration human interaction, electricity and clean air legislation can be identified as the major causes of climate change. Computer analysis and Internet misinformation contributed to the rise of junk science and diverted attention away from actual solutions. Acceptance of these findings will depend of individual more than government.

What Now?

Dealing with the effects of clean air legislation on the environment will be like unraveling a skein of yarn that a cat has played with for fifty years. Trying to figure out a way to effectively incorporate electricity into an environment that survives on air pollution is not as simple as going off the grid. Fossil fuels will need to be part of the solution but not the only part. Moderation is the key here and this is not a battle for taking back the atmosphere. The world cannot afford collateral damage this time around.

Government cannot and, should not, be tasked with solving this problem alone. The issues of unsustainable cities and rural communities that bear too much responsibility for a healthy environment have to be addressed. Chasing a one-shot, solve everything method of maintaining the status quo is not in the cards. Multiple perspectives are needed and there is no room for armchair quarterbacks at the table. While it may sound cliche and trite, these are phrases used for the purpose of making people understand this situation is serious. Climate change is real. Scientific warnings are based on essentially accurate data even if they draw inaccurate conclusions. There is no moral ground here that makes one perspective more valuable than another, only the need for understanding and commitment as well as flexibility.

Most of all, recovery will take time and patience. Three generations have had their health jeopardized by an absence of sulfur in the atmosphere. Wildlife has been endangered by the same deficiency. It will take a slow and calculated response to make this happen but seeing a return to normal in as little as five years is possible. The environment is more resilient than believed and when it has the "food" it needs, its health will return as well as the world's.

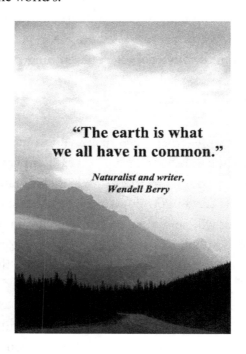

Sources

Special Note from the Author:

This book was based on ten years of random late night Internet research and six decades of life experience. If every tidbit of information put in this book was cited, these source pages would be longer than the book itself. Rather than pretend to be the academic she is not, the author has chosen to provide a list of websites instead of specific pages. Her definition of a factual cite was two-fold. First if the reference corresponded or backed up to her own memory, it was used to support opinion statements. Second, the scientific information had to match what she had been taught prior to Internet use and/or multiple websites without copy/pasted verbiage had to be found. Those may not be accepted standards for citing books such as this but with so much plaigeraized and misquoted information on the internet, there needed to be some type of criteria to support a theory that runs counter to popular culture.

- **Wikipedia:** Not considered a quality literary source, publishing needs to recognized how far this source has come. Representing what the internet was supposed to be, Wikipedia has shunned the money making aspect of the internet and allowed individuals from all backgrounds to offer information that might be helpful to someone. As it has tightened its guidelines and operation, it has become one of the best ways to quickly check a date or find a list that would not be accessible in more specialized webpages.

- **United States Government**: In questioning the actions of the federal government, it was best to use their own information as a primary source. The number of webpages that were accessed is hard to ascertain. A list of all the main department websites that were used follows.
    - **Environmental Protection Agency** (www.epa.gov)
    - **National Aeronautics and Space Administration NASA** (climatekids.nasa.gov), (earthobservatory.nasa.gov)
    - **National Oceanic and Atmospheric Administration** (www.noaa.gov) (www.weather.gov)
    - **Department of Agriculture** (naldc.nal.usda.gov) as well as other State Cooperative Extension agencies
    - **National Institute of Health: Digital Collection for Biological Information** (www.ncbi.nlm.nih.gov)
    - **Department of the Interior** (www.doi.gov)
- **Encyclopedia Britannica** (www.britannica.com) *One of the best encylcopedia sources before the digital age and one that is working to keep information freely available.
- **History.com** (www.history.com) As long as the user remembers that this is a companion site to the cable channel of a similar name, the information can provides clues as to how historical events coincide.
- Science Direct (www.sciencedirect.com) is a peer reviewed collection of papers and educational information. Hosted by Elsevier in a generally open document format, this globally curated and reviewed collection of science includes more than one million titles ranging from basic principles to technical explanations.
- California Based Websites: Information about California and its environmental struggles is readily available but specific. Websites with the California state address (www.ca.gov) as well as the individual websites of large cities and those recording historic events.

Photo Credits- Photos used here are believed to be public domain files or cited in the caption.

Informational Graphics

There is no denying that a graph or chart can illustrate a mathematical relationship much better than simple numbers. This section will provide those pictorial representations. Each image also includes a summary of the information to which is refers.

Carbon Monoxide (1980-2019)

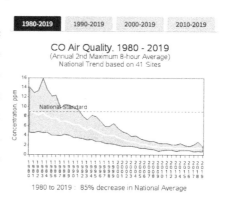

Carbon monoxide might be dangerous to animal life in concentrations of more than 10% of available air but it may also be a valuable component when this gas is free circulating. Today, carbon based gases are estimated at less than 1% of the atmosphere and there are calls to reduce their presence even more. This average comes from 41 location, less than one per state, and given the ever changing weather, may not be a fair representation of the air quality in a country as diverse as the United States. Can vegetation survive on such low levels of carbon based gases.

**Note: CO levels have been less than the national standard since the mid 1990s.

Nitrogen Trends (1980-2019)

Nitrogen is the most common element in the atmosphere. It provides a safe mixer for oxygen and carbon dioxide used respectively by animals and plants. Both are dangerous in higher levels. Without nitrogen, the air would be too rich in oxygen and too deadly with carbon dioxide. Yet, the EPA regulates nitrogen dioxide. Beneficial to 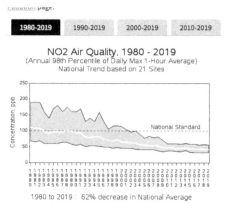 agriculture and natural habitats, nature sees that NO2 is formed by lightning activity during thunderstorms and geological events. This average takes samples from only 21 locations and shows a 62% reduction since 1980. Nitrogen oxides are also breathable forms of nitrogen when freely circulating and may provide levels of nitrogen which help the human body maintain protein levels and a pleasant outlook.

Ozone Trends (1980-2019)

Taken from the EPA website, this graph shows a 35% reduction in ozone levels since 1980. Keep in mind that the graph shows a stable National standard throughout these forty years but shows no indication that the standard has been increased to accommodate the needs of the additional 100 million 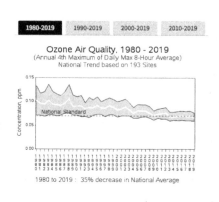 people who now call the United States home. Of the available air quality charts available on the EPA website this samples the largest number of location but still relatively low for a country that includes over 3.5 million square miles of space.

Sulfur Trends (1980-2019)

Of all the gases labeled pollutants, targeting sulfur dioxide may have been the most disruptive to the environment. Needed by both plants and animals, the EPA has noted a 92% decline in sulfur dioxide levels since 1980. The compound is breathable in small quantities such as in freely circulated air.

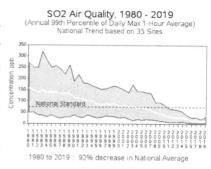

It may be possible for the human body to utilize sulfur from the air to support the production of insulin, more flexible blood vessels and a increase and maintain cartilage in the knees and spine. Health benefits of sulfur are well documented. As sulfure dioxide in the atmosphere has declined, the incidence of high blood pressure, arthritis and diabetes has increased. Sampled in only 35 sites, this estimation of sulfur levels may be greatly over reported given the need for this element in maintaining wildlife habitats and agricultural lands.

To better represent the decline in atmospheric sulfur for farm use, the following maps indicate the decline of sulfur over a much smaller period of time. The land in the darker reds is prime farm land which surrounds heavily industrialized locations. Now those areas have a fraction of available sulfur which weakens the food supply and contributes to human health problems. (The original size of this photo prohibited its enlargement)

Particulates (1990-2019)

Particulate matter is vital to the formation of clouds that bring rain. The EPA tracks two different levels but only one is shown here. The second category is for sumaller particles and has been consistently below the national standard since records began. The average is based on 111 locations. It should be noted this chart would likely include measuring ash produced in volcanic eruptions and other natural events such as wild fires.

Region 9 EPA Trends of Sulfur Dioxide from 1970-2018

While most of the charts posted here are for periods from 1980 to Present, this chart perhaps gives a more complete picture of how close the country is to zero tolerance of gases listed by the EPA as pollutants.

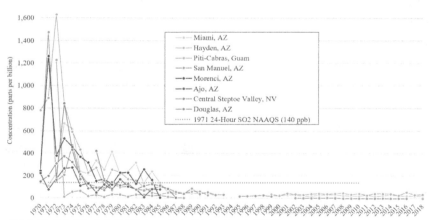

*As designated under the 1971 sulfur dioxide (SO₂) primary national ambient air quality standard (NAAQS).
Source: US EPA's Air Quality Systems (AQS) database (July 18, 2019). The 1971 NAAQS for 24-hour SO₂ is 140 parts per billion (ppb), not to be exceeded more than once in a calendar year. The form of the design value for primary 24-hour SO₂ is the second-highest average value, based on hourly data that are at least 75% complete for each calendar quarter. The primary 24-hour SO₂ standard was revoked in 2010. No monitoring data is available for the *Piti-Cabras* nonattainment area in Guam or the *Coconino County, AZ* unclassifiable area.
AIR19100 - 2018 annual air quality update - SO2.xlsx (November 27, 2019)

Methionine Sources

While food lists vary by website, this chart shows the type of inaccurate nutritional information often found on the Internet. Methionine is an amino acid that can be found in animal and plant proteins. However, it is not the only amino acid present at one time. More importantly, RDIs (Recommended Daily Intake) for amino acids are not generally provided. In fact, RDIs for proteins have largely been abandoned as a result of individuals being unable to maintain muscle mass under stricter dietary guidelines. It should be noted that like plants, animals (including fish) need to have adequate access to sulfur for methionine to be produced. The inclusion on methionine in a diet does not ensure adequate levels of sulfur for metabolism. RDIs are generally based on a 2000 calorie diet rather than individual needs.

About the Author

Describing her life to date as eclectic, Sarah Schrumpf-Deacon has more than six decades in the rear view mirror when it comes to life on planet Earth. Born in a small lumber and railroad town that included a well-respected Presbyterian college, she would reflect on how these early years would be a glimpse of what her life would hold.

Farming, forestry and country living were to become as important to her perspective as being raised in a large growing manufacturing community with its art, music and academic offerings. A college education in what was then known as Home Economics taught her how interconnected life was. From public health and family financial wellness to nutrition and political science, her love of learning only expanded as she tackled animal nutrition, soil testing and plant propagation. Farm ownership was one of her greatest joys and challenges. Twenty-five years of producing food for local markets, she understands that the first step to ensuring good nutrition for the public comes from a well maintained environment, not a doctor's prescription pad. It is this diverse background that led to the writing of **Air Pollution's the Answer!**

Not a reader of fantasy but intrigued by historical non-fiction titles, her literary preference runs to biographies and historical fiction. Biographies of women like Queen Victoria, Mary Todd Lincoln, Helen Keller, Marie Curie and Pearl S. Buck offered proof of how smart and influential women have been throughout history, even in traditional and socially restrictive roles. Comparative works like Lee and Grant by Gene Smith added a personal perspective and back story as to how and why critically important historical events were put into motion. James A. Michener's

Hawaii and similar works, detailed how social conflict often finds it roots in environmental issues. For a change of pace, a good science based mystery like **The Body Farm** offered her factual knowledge in an entertaining context.

Always wanting to write professionally, an opportunity to contribute to a local newspaper came in 2000, after farming and teaching for several years. With an eleven year return to education between then and now, she finally calls freelance writing her job. Never expecting to find herself the author of a book on climate change, the nutrition advice she planned turned into so much more.

Sarah Schrumpf-Deacon continues to accept freelance writing jobs but enjoys semi-retired life in rural Virginia with her husband and dog. She maintains a writer's blog called ***Just a Touch of Sass*** (www.justatouchofsass.com). Her Nana Jane's Words/Books handle was the creation of her three, now grown and successful, children over fifteen years ago. Hating the name for how she wanted to be addressed as a grandmother, it became the perfect persona for direct and often blunt advice and observations now posted on her blog.

CPSIA information can be obtained
at www.ICGtesting.com
Printed in the USA
BVHW050439200523
664488BV00006B/799